Collins

2020 GUIDE
to the
NIGHT SKY

Storm Dunlop and Wil Tirion

Published by Collins
An imprint of HarperCollins Publishers
Westerhill Road
Bishopbriggs
Glasgow G64 2QT
www.harpercollins.co.uk

In association with
Royal Museums Greenwich, the group name for the National Maritime Museum,
Royal Observatory Greenwich, Queen's House and *Cutty Sark*
www.rmg.co.uk

© HarperCollins Publishers 2019
Text and illustrations © Storm Dunlop and Wil Tirion
Photographs © see acknowledgements page 110

A catalogue record for this book is available from the British Library

ISBN 978-0-00-825771-2

10 9 8 7 6 5 4 3 2

Printed in Great Britain by Bell and Bain Ltd, Glasgow

If you would like to comment on any aspect of this book, please contact us at the above address or online.
e-mail: collinsmaps@harpercollins.co.uk

 facebook.com/CollinsAstronomy

@CollinsAstro

Contents

Introduction

The aim of this Guide is to help people find their way around the night sky, by showing how the stars that are visible change from month to month and by including details of various events that occur throughout the year. The objects and events described may be observed with the naked eye, or nothing more complicated than a pair of binoculars.

The conditions for observing naturally vary over the course of the year. During the summer, twilight may persist throughout the night and make it difficult to see the faintest stars. There are three recognized stages of twilight: civil twilight, when the Sun is less than 6° below the horizon; nautical twilight, when the Sun is between 6° and 12° below the horizon; and astronomical twilight, when the Sun is between 12° and 18° below the horizon. Full darkness occurs only when the Sun is more than 18° below the horizon. During nautical twilight, only the very brightest stars are visible. During astronomical twilight, the faintest stars visible to the naked eye may be seen directly overhead, but are lost at lower altitudes. As the diagram shows, during June and most of July full darkness never occurs at the latitude of London, and at Edinburgh nautical twilight persists throughout the whole night, so at that latitude only the very brightest stars are visible.

Another factor that affects the visibility of objects is the amount of moonlight in the sky. At Full Moon, it may be very difficult to see some of the fainter stars and objects, and even when the Moon is at a smaller phase it may seriously interfere with visibility if it is near the stars or planets in which you are interested. A full lunar calendar is given for each month and may be used to see when nights are likely to be darkest and best for observation.

The celestial sphere

All the objects in the sky (including the Sun, Moon and stars) appear to lie at some indeterminate distance on a large sphere, centred on the Earth. This *celestial sphere* has various reference points and features that are related to those of the Earth. If the Earth's rotational axis is extended, for example, it points to the North and South Celestial Poles, which are thus in line with the North and South Poles on Earth. Similarly, the *celestial*

The duration of twilight throughout the year at London and Edinburgh.

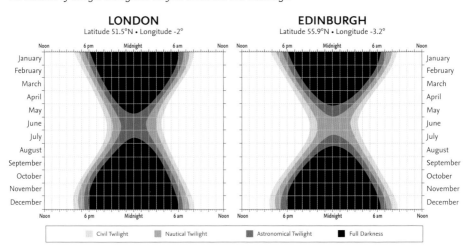

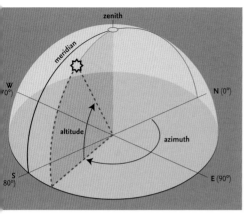

Measuring altitude and azimuth on the celestial sphere.

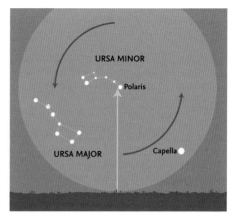

The altitude of the North Celestial Pole equals the observer's latitude.

equator lies in the same plane as the Earth's equator, and divides the sky into northern and southern hemispheres. Because this Guide is written for use in Britain and Ireland, the area of the sky that it describes includes the whole of the northern celestial hemisphere and those portions of the southern that become visible at different times of the year. Stars in the far south, however, remain invisible throughout the year, and are not included.

It is useful to know some of the special terms for various parts of the sky. As seen by an observer, half of the celestial sphere is invisible, below the horizon. The point directly overhead is known as the **zenith**, and the (invisible) one below one's feet as the **nadir**. The line running from the north point on the horizon, up through the zenith and then down to the south point is the **meridian**. This is an important invisible line in the sky, because objects are highest in the sky, and thus easiest to see, when they cross the meridian in the south. Objects are said to **transit**, when they cross this line in the sky.

In this book, reference is frequently made in the text and in the diagrams to the standard compass points around the horizon. The position of any object in the sky may be described by its **altitude** (measured in degrees above the horizon), and its **azimuth** (measured in degrees from north 0°, through east 90°,

south 180° and west 270°). Experienced amateurs and professional astronomers also use another system of specifying locations on the celestial sphere, but that need not concern us here, where the simpler method will suffice.

The celestial sphere appears to rotate about an invisible axis, running between the North and South Celestial Poles. The location (i.e., the altitude) of the Celestial Poles depends entirely on the observer's position on Earth or, more specifically, their latitude. The charts in this book are produced for the latitude of 50°N, so the North Celestial Pole (NCP) is 50° above the northern horizon. The fact that the NCP is fixed relative to the horizon means that all the stars within 50° of the Pole are always above the horizon and may, therefore, always be seen at night, regardless of the time of year. This northern circumpolar region is an ideal place to begin learning the sky, and ways to identify the circumpolar stars and constellations will be described shortly.

The ecliptic and the zodiac

Another important line on the celestial sphere is the Sun's apparent path against the background stars – in reality the result of the Earth's orbit around the Sun. This is known as the **ecliptic**. The point where the Sun, apparently moving along the ecliptic, crosses the celestial equator from south to north is

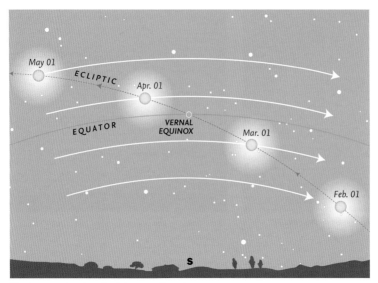

The Sun crossing the celestial equator in spring.

known as the vernal (or spring) *equinox*, which occurs on March 20. At this time (and at the autumnal equinox, on September 22 or 23, when the Sun crosses the celestial equator from north to south) day and night are almost exactly equal in length. (There is a slight difference, but that need not concern us here.) The vernal equinox is currently located in the constellation of Pisces, and is important in astronomy because it defines the zero point for a system of celestial coordinates, which is, however, not used in this Guide.

The Moon and planets are to be found in a band of sky that extends 8° on either side of the ecliptic. This is because the orbits of the Moon and planets are inclined at various angles to the ecliptic (i.e., to the plane of the Earth's orbit). This band of sky is known as the *zodiac* and, when originally devised, consisted of twelve *constellations*, all of which were considered to be exactly 30° wide. When the constellation boundaries were formally established by the International Astronomical Union in 1930, the exact extent of most constellations was altered and, nowadays, the ecliptic passes through thirteen constellations.

Because of the boundary changes, the Moon and planets may actually pass through several other constellations that are adjacent to the original twelve.

The constellations

Since ancient times, the celestial sphere has been divided into various constellations, most dating back to antiquity and usually associated with certain myths or legendary people and animals. Nowadays, the boundaries of the constellations have been fixed by international agreement and their names (in Latin) are largely derived from Greek or Roman originals. Some of the names of the most prominent stars are of Greek or Roman origin, but many are derived from Arabic names. Many bright stars have no individual names and, for many years, stars were identified by terms such as 'the star in Hercules' right foot'. A more sensible scheme was introduced by the German astronomer Johannes Bayer in the early seventeenth century. Following his scheme – which is still used today – most of the brightest stars are identified by a Greek letter followed by the genitive form of the constellation's Latin name. An example is the Pole Star, also known as Polaris and α Ursae Minoris (abbreviated α UMi). The Greek alphabet is shown on page 109 with a list of all the constellations that may be seen from latitude 50°N, together with abbreviations, their genitive forms and English names. Other naming schemes exist for fainter stars, but are not used in this book.

Asterisms

Apart from the constellations (88 of which cover the whole sky), certain groups of stars, which may form a part of a larger constellation or cross several constellations, are readily recognizable and have been given individual names. These groups are known as *asterisms*, and the most famous (and well-known) is the 'Plough', the common name for the seven brightest stars in the constellation of Ursa Major, the Great Bear. The names and details of some asterisms mentioned in this book are given in the list on page 110.

Magnitudes

The brightness of a star, planet or other body is frequently given in *magnitudes* (mag.). This is a mathematically defined scale where larger numbers indicate a fainter object. The scale extends beyond the zero point to negative numbers for very bright objects. (Sirius, the brightest star in the sky is mag. -1.4.) Most observers are able to see stars to about mag. 6, under very clear skies.

The Moon

Although the daily rotation of the Earth carries the sky from east to west, the Moon gradually moves eastwards by approximately its diameter (about half a degree) in an hour. Normally, in its orbit around the Earth, the Moon passes above or below the direct line between Earth and Sun (at New Moon) or outside the area obscured by the Earth's shadow (at Full Moon). Occasionally, however, the three bodies are more-or-less perfectly aligned to give an *eclipse*: a solar eclipse at New Moon or a lunar eclipse at Full Moon. Depending on the exact circumstances, a solar eclipse may be merely partial (when the Moon does not cover the whole of the Sun's disk); annular (when the Moon is too far from Earth in its orbit to appear large enough to hide the whole of the Sun); or total. Total and annular eclipses are visible from very restricted areas of the Earth, but partial eclipses are normally visible over a wider area.

Somewhat similarly, at a lunar eclipse, the Moon may pass through the outer zone of the Earth's shadow, the *penumbra* (in a penumbral eclipse, which is not generally perceptible to the naked eye), so that just part of the Moon is within the darkest part of the Earth's shadow, the *umbra* (in a partial eclipse); or completely within the umbra (in a total eclipse). Unlike solar eclipses, lunar eclipses are visible from large areas of the Earth.

Occasionally, as it moves across the sky, the Moon passes between the Earth and individual planets or distant stars, giving rise to an *occultation*. As with solar eclipses, such occultations are visible from restricted areas of the world.

The planets

Because the planets are always moving against the background stars, they are treated in some detail in the monthly pages and information is given when they are close to other planets, the Moon or any of five bright stars that lie near the ecliptic. Such events are known as *appulses* or, more frequently, as *conjunctions*. (There are technical differences in the way these terms are defined – and should be used – in astronomy, but these need not concern us here.) The positions of the planets are shown for every month on a special chart of the ecliptic.

The term conjunction is also used when a planet is either directly behind or in front of the Sun, as seen from Earth. (Under normal circumstances it will then be invisible.) The conditions of most favourable visibility depend on whether the planet is one of the two known as *inferior planets* (Mercury and Venus) or one of the three *superior planets* (Mars, Jupiter and

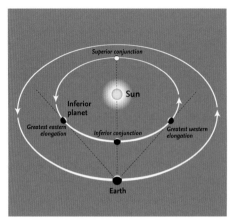

Inferior planet.

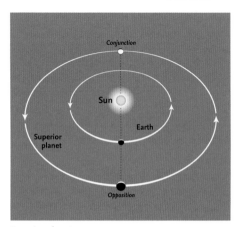

Superior planet.

Saturn) that are covered in detail. (Some details of the fainter superior planets, Uranus and Neptune, are included in this Guide, and special charts are given on page 25.)

The inferior planets are most readily seen at eastern or western **elongation**, when their angular distance from the Sun is greatest. For superior planets, they are best seen at **opposition**, when they are directly opposite the Sun in the sky, and cross the meridian at local midnight.

It is often useful to be able to estimate angles on the sky, and approximate values may be obtained by holding one hand at arm's length. The various angles are shown in the diagram, together with the separations of the various stars in the Plough.

Meteors

At some time or other, nearly everyone has seen a **meteor** – a 'shooting star' – as it flashed across the sky. The particles that cause meteors – known technically as 'meteoroids' – range in size from that of a grain of sand (or even smaller) to the size of a pea. On any night of the year there are occasional meteors, known as **sporadics**, that may travel in any direction. These occur at a rate that is normally between three and eight in an hour. Far more important, however, are **meteor showers**, which occur at fixed periods of the year, when the

Earth encounters a trail of particles left behind by a comet or, very occasionally, by a minor planet (asteroid). Meteors always appear to diverge from a single point on the sky, known as the **radiant**, and the radiants of major showers are shown on the monthly charts. Meteors that come from a circular area 8° in diameter around the radiant are classed as belonging to the particular shower. All others that do not come from that area are sporadics (or occasionally from another shower that is active at the same time). A list of the major meteor showers is given on page 31.

Although the positions of the various shower radiants are shown on the monthly charts, looking directly at the radiant is not the most effective way of seeing meteors. They are most likely to be noticed if one is looking about 40–45° away from the radiant position. (This is approximately two hand-spans as shown in the diagram for measuring angles.)

Other objects

Certain other objects may be seen with the naked eye under good conditions. Some were given names in antiquity – Praesepe is one example – but many are known by what are called 'Messier numbers', the numbers in a catalogue of nebulous objects compiled by Charles Messier in the late eighteenth century. Some, such as the Andromeda Galaxy, M31, and the Orion Nebula, M42, may be seen

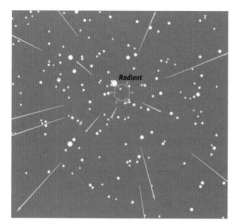

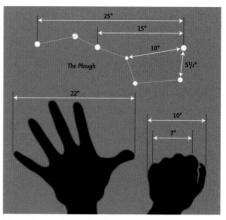

Meteor shower (showing the April Lyrid radiant).

Measuring angles in the sky.

by the naked eye, but all those given in the list will benefit from the use of binoculars. Apart from galaxies, such as M31, which contain thousands of millions of stars, there are also two types of cluster: open clusters, such as M45, the Pleiades, which may consist of a few dozen to some hundreds of stars; and globular clusters, such as M13 in Hercules, which are spherical concentrations of many thousands of stars. One or two gaseous nebulae, consisting of gas illuminated by stars within them, are also visible. The Orion Nebula, M42, is one, and is illuminated by the group of four stars, known as the Trapezium, which may be seen within it by using a good pair of binoculars.

Some interesting objects.

Messier / NGC	Name	Type	Constellation	Maps (months)
—	Hyades	open cluster	Taurus	Sep. – Apr.
—	Double Cluster	open cluster	Perseus	All year
—	Melotte 111 (Coma Cluster)	open cluster	Coma Berenices	Jan. – Aug.
M3	—	globular cluster	Canes Venatici	Jan. – Sep.
M4	—	globular cluster	Scorpius	May – Aug.
M8	Lagoon Nebula	gaseous nebula	Sagittarius	Jun. – Sep.
M11	Wild Duck Cluster	open cluster	Scutum	May – Oct.
M13	Hercules Cluster	globular cluster	Hercules	Feb. – Nov.
M15	—	globular cluster	Pegasus	Jun. – Dec.
M22	—	globular cluster	Sagittarius	Jun. – Sep.
M27	Dumbbell Nebula	planetary nebula	Vulpecula	May – Dec.
M31	Andromeda Galaxy	galaxy	Andromeda	All year
M35	—	open cluster	Gemini	Oct. – May
M42	Orion Nebula	gaseous nebula	Orion	Nov. – Mar.
M44	Praesepe	open cluster	Cancer	Nov. – Jun.
M45	Pleiades	open cluster	Taurus	Aug. – Apr.
M57	Ring Nebula	planetary nebula	Lyra	Apr. – Dec.
M67	—	open cluster	Cancer	Dec. – May
NGC 752	—	open cluster	Andromeda	Jul. – Mar.
NGC 3242	Ghost of Jupiter	planetary nebula	Hydra	Feb. – May

The Northern Circumpolar Constellations

The northern circumpolar stars are the key to starting to identify the constellations. For anyone in the northern hemisphere they are visible at any time of the year, and nearly everyone is familiar with the seven stars of the Plough – known as the Big Dipper in North America – an asterism that forms part of the large constellation of **Ursa Major** (the Great Bear).

Ursa Major

Because of the movement of the stars caused by the passage of the seasons, Ursa Major lies in different parts of the evening sky at different periods of the year. The diagram below shows its position for the four main seasons. The seven stars of the Plough remain visible throughout the year anywhere north of latitude 40°N. Even at the latitude (50°N) for which the charts in this book are drawn, many of the stars in the southern portion of the constellation of Ursa Major are hidden below the horizon for part of the year or (particularly in late summer) cannot be seen late in the night.

Polaris and Ursa Minor

The two stars **Dubhe** and **Merak** (α and β Ursae

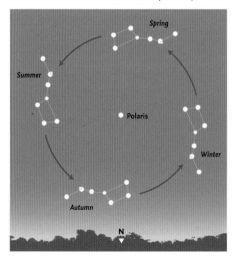

Majoris, respectively), farthest from the 'tail' are known as the 'Pointers'. A line from Merak to Dubhe, extended about five times their separation, leads to the Pole Star, **Polaris**, or α Ursae Minoris. All the stars in the northern sky appear to rotate around it. There are five main stars in the constellation of **Ursa Minor**, and the two farthest from the Pole, **Kochab** and **Pherkad** (β and γ Ursae Minoris, respectively), are known as 'The Guards'.

Cassiopeia

On the opposite of the North Pole from Ursa Major lies **Cassiopeia**. It is highly distinctive, appearing as five stars forming a letter 'W' or 'M' depending on its orientation. Provided the sky is reasonably clear of clouds, you will nearly always be able to see either Ursa Major or Cassiopeia, and thus be able to orientate yourself on the sky.

To find Cassiopeia, start with **Alioth** (ε Ursae Majoris), the first star in the tail of the Great Bear. A line from this star extended through Polaris points directly towards γ Cassiopeiae, the central star of the five.

Cepheus

Although the constellation of **Cepheus** is fully circumpolar, it is not nearly as well known as Ursa Major, Ursa Minor or Cassiopeia, partly because its stars are fainter. Its shape is rather like the gable end of a house. The line from the Pointers through Polaris, if extended, leads to **Errai** (γ Cephei) at the 'top' of the 'gable'. The brightest star, **Alderamin** (α Cephei) lies in the Milky Way region, at the 'bottom right-hand corner' of the figure.

Draco

The constellation of **Draco** consists of a quadrilateral of stars, known as the 'Head of Draco' (and also the 'Lozenge'), and a long chain of stars forming the neck and body of the dragon. To find the Head of Draco, locate the two stars **Phecda** and **Megrez** (γ and δ Ursae Majoris) in the Plough, opposite the Pointers.

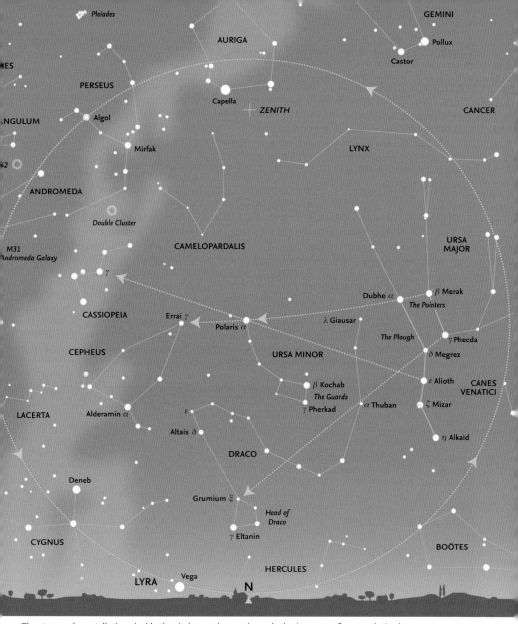

The stars and constellations inside the circle are always above the horizon, seen from our latitude.

Extend a line from Phecda through Megrez by about eight times their separation, right across the sky below the Guards in Ursa Minor, ending at **Grumium** (ξ Draconis) at one corner of the quadrilateral. The brightest star, **Eltanin** (γ Draconis) lies farther to the south. From the head of Draco, the constellation first runs northeast to **Altais** (δ Draconis) and ε Draconis, then doubles back southwards before winding its way through **Thuban** (α Draconis) before ending at **Giausar** (λ Draconis) between the Pointers and Polaris.

The Winter Constellations

The winter sky is dominated by several bright stars and distinctive constellations. The most conspicuous constellation is **Orion**, the main body of which has an hourglass shape. It straddles the celestial equator and is thus visible from anywhere in the world. The three stars that form the 'Belt' of Orion point down towards the southeast and to **Sirius** (α Canis Majoris), the brightest star in the sky. **Mintaka** (δ Orionis), the star at the northeastern end of the Belt, farthest from Sirius, actually lies just slightly south of the celestial equator.

A line from **Bellatrix** (γ Orionis) at the 'top right-hand corner' of Orion, through **Aldebaran** (α Tauri), past the 'V' of the Hyades cluster, points to the distinctive cluster of bright blue stars known as the **Pleiades**, or the 'Seven Sisters'. Aldebaran is one of the five bright stars that may sometimes be occulted (hidden) by the Moon. Another line from Bellatrix, through **Betelgeuse** (α Orionis), if carried right across the sky, points to the constellation of **Leo**, a prominent constellation in the spring sky.

Six bright stars in six different constellations: **Capella** (α Aurigae), **Aldebaran** (α Tauri), **Rigel** (β Orionis), **Sirius** (α Canis Majoris), **Procyon** (α Canis Minoris) and **Pollux** (β Gemini) form what is sometimes known as the 'Winter Hexagon'. Pollux is accompanied to the northwest by the slightly fainter star of **Castor** (α Gemini), the second 'Twin'.

In a counterpart to the famous 'Summer Triangle', an almost perfect equilateral triangle, the 'Winter Triangle', is formed by Betelgeuse (α Orionis), Sirius (α Canis Majoris) and Procyon (α Canis Minoris).

Several of the stars in this region of the sky show distinctive tints: Betelgeuse (α Orionis) is reddish, Aldebaran (α Tauri) is orange, and Rigel (β Orionis) is blue-white.

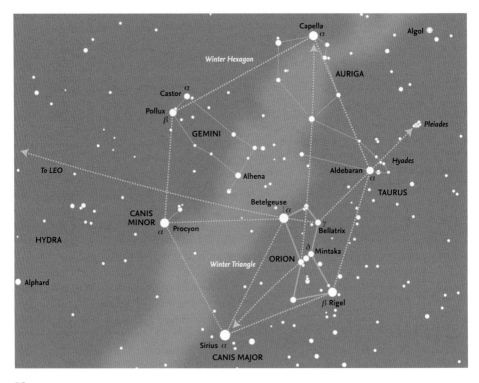

The Spring Constellations

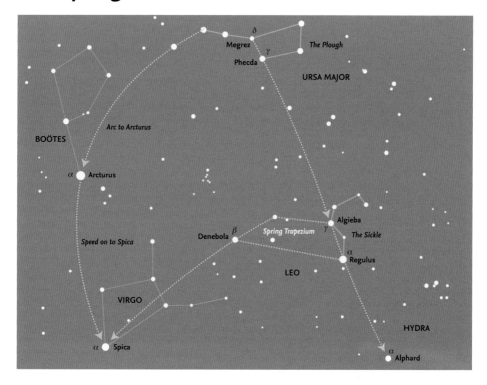

The most prominent constellation in the spring sky is the zodiacal constellation of **Leo**, and its brightest star, **Regulus** (α Leonis), which may be found by extending a line from Megrez and Phecda (δ and γ Ursae Majoris, respectively) – the two stars on the opposite side of the bowl of the Plough from the Pointers – down to the southeast. Regulus forms the 'dot' of the 'backward question mark' known as 'the Sickle'. Regulus, like Aldebaran in Taurus, is one of the bright stars that lie close to the ecliptic, and that are occasionally occulted by the Moon. The same line from Ursa Major to Regulus, if continued, leads to **Alphard** (α Hydrae), the brightest star in **Hydra**, the largest of the 88 constellations.

The shape formed by the body of Leo is sometimes known as the 'Spring Trapezium'. At the other end of the constellation from Regulus is **Denebola** (β Leonis), and the line forming the back of the constellation through Denebola points to the bright star **Spica** (α Virginis) in the constellation of **Virgo**. A saying that helps to locate Spica is well known to astronomers: 'Arc to Arcturus and then speed on to Spica.' This suggests following the arc of the tail of Ursa Major to Arcturus and then on to Spica. **Arcturus** (α Boötis) is actually the brightest star in the northern hemisphere of the sky. (Although other stars, such as Sirius, are brighter, they are all in the southern hemisphere.) Overall, the constellation of **Boötes** is sometimes described as 'kite-shaped' or 'shaped like the letter P'.

Although Spica is the brightest star in the zodiacal constellation of Virgo, the rest of the constellation is not particularly distinct, consisting of a rough quadrilateral of moderately bright stars and some fainter lines of stars extending outwards.

The Summer Constellations

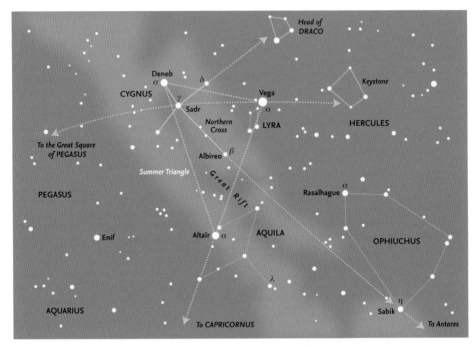

On summer nights, the three bright stars **Deneb** (α Cygni), **Vega** (α Lyrae) and **Altair** (α Aquilae) form the striking 'Summer Triangle'. The constellations of **Cygnus** (the Swan) and **Aquila** (the Eagle) represent birds 'flying' down the length of the Milky Way. This part of the Milky Way contains the **Great Rift**, an elongated dark region, where the light from distant stars is obscured by intervening dust. The dark Rift is clearly visible even to the naked eye.

The most prominent stars of Cygnus are sometimes known as the 'Northern Cross' (as a counterpart to the 'Southern Cross' – the constellation of Crux – in the southern hemisphere). The central line of Cygnus through **Albireo** (β Cygni), extended well to the southwest, points to **Sabik** (η Ophiuchi) in the large, sprawling constellation of Ophiuchus (the Serpent Bearer) and beyond to **Antares** (α Scorpii) in the constellation of **Scorpius**. Like Cepheus, the shape of Ophiuchus somewhat resembles the gable-end of a house, and the

brightest star **Rasalhague** (α Ophiuchi) is at the 'apex' of the 'gable'.

A line from the central star of Cygnus, **Sadr** (γ Cygni) through δ Cygni, in the northwestern 'wing' points towards the Head of Draco, and is another way of locating that part of the constellation. Another line from Sadr to Vega indicates the central portion, 'The Keystone', of the constellation of **Hercules**. An arc through the same stars, in the opposite direction, points towards the constellation of **Pegasus**, and more specifically to the 'Great Square of Pegasus'.

Aquila is less conspicuous than Cygnus and consists of a diamond shape of stars, representing the body and wings of the eagle, together with a rather faint star, λ Aquilae, marking the 'head'. **Lyra** (the Lyre) mainly consists of Vega (α Lyrae) and a small quadrilateral of stars to its southeast. Continuation of a line from Vega through Altair indicates the zodiacal constellation of **Capricornus**.

The Autumn Constellations

During the autumn season, the most striking feature is the 'Great Square of Pegasus', an almost perfect rectangle on the sky, forming the main body of the constellation of **Pegasus**. However, the star at the northeastern corner, **Alpheratz**, is actually α Andromedae, and part of the adjacent constellation of **Andromeda**. A line from **Scheat** (β Pegasi) at the northeastern corner of the Square, through **Matar** (η Pegasi), points in the general direction of Cygnus. A crooked line of stars leads from **Markab** (α Pegasi) through **Homam** (ζ Pegasi) and **Biham** (θ Pegasi) to **Enif** (ε Pegasi). A line from Markab through the last star in the Square, **Algenib** (γ Pegasi) points in the general direction of the five stars, including **Menkar** (α Ceti) that form the 'tail' of the constellation of **Cetus** (the Whale). A ring of seven stars lying below the southern side of the Great Square is known as 'the Circlet', part of the constellation of **Pisces** (the Fishes).

Extending the line of the western side of the Great Square towards the south leads to the isolated bright star, **Fomalhaut** (α Piscis Austrini), in the Southern Fish. Following the line of the eastern side of the Great Square towards the north leads to Cassiopeia while, in the other direction, it points towards **Diphda** (β Ceti), which is actually the brightest star in Cetus.

Three bright stars leading northeast from Alpheratz form the main body of the constellation of **Andromeda**. Continuation of that line leads towards the constellation of **Perseus** and **Mirfak** (α Persei). Running southwards from Mirfak is a chain of stars, one of which is the famous variable star **Algol** (β Persei). Farther east, an arc of stars leads to the prominent cluster of the **Pleiades**, in the constellation of **Taurus.**

Between Andromeda and Cetus lie the two small constellations of **Triangulum** (the Triangle) and **Aries** (the Ram).

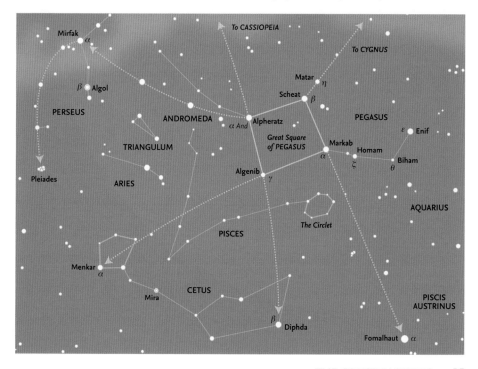

The Moon at First Quarter.

The Moon

The monthly pages include diagrams showing the phase of the Moon for every day of the month, and also indicate the day in the *lunation* (or *age* of the Moon), which begins at New Moon. Although the main features of the surface – the light highlands and the dark maria (seas) – may be seen with the naked eye, far more features may be detected with the use of binoculars or any telescope. The many craters are best seen when they are close to the *terminator* (the boundary between the illuminated and the non-illuminated areas of the surface), when the Sun rises or sets over any particular region of the Moon and the crater walls or central peaks cast strong shadows. Most features become difficult to see at Full Moon, although this is the best time to see the bright ray systems surrounding certain craters. Accompanying the Moon map on the following pages is a list of prominent features, including the days in the lunation when features are normally close to the terminator and thus easiest to see. A few bright features

such as Linné and Proclus, visible when well illuminated, are also listed. One feature, Rupes Recta (the Straight Wall) is readily visible only when it casts a shadow with light from the east, appearing as a light line when illuminated from the opposite direction.

The dates of visibility vary slightly through the effects of *libration*. Because the Moon's orbit is inclined to the Earth's equator and also because it moves in an ellipse, the Moon appears to rock slightly from side to side (and nod up and down). Features near the *limb* (the edge of the Moon) may vary considerably in their location and visibility. (This is easily noticeable with Mare Crisium and the craters Tycho and Plato.) Another effect is that at crescent phases before and after New Moon, the normally non-illuminated portion of the Moon receives a certain amount of light, reflected from the Earth. This *Earthshine* may enable certain bright features (such as Aristarchus, Kepler and Copernicus) to be detected even though they are not illuminated by sunlight.

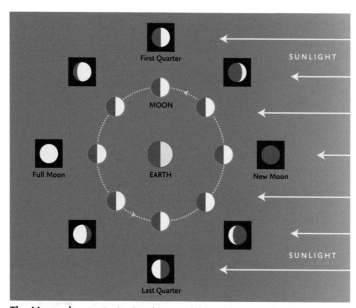

The Moon phases. *During its orbit around the Earth we see different portions of the illuminated side of the Moon's surface.*

Map of the Moon

Abulfeda	6:20
Agrippa	7:21
Albategnius	7:21
Aliacensis	7:21
Alphonsus	8:22
Anaxagoras	9:23
Anaximenes	11:25
Archimedes	8:22
Aristarchus	11:25
Aristillus	7:21
Aristoteles	6:20
Arzachel	8:22
Atlas	4:18
Autolycus	7:21
Barrow	7:21
Billy	12:26
Birt	8:22
Blancanus	9:23
Bullialdus	9:23
Bürg	5:19
Campanus	10:24
Cassini	7:21
Catharina	6:20
Clavius	9:23
Cleomedes	3:17
Copernicus	9:23
Cyrillus	6:20
Delambre	6:20
Deslandres	8:22
Endymion	3:17
Eratosthenes	8:22
Eudoxus	6:20
Fracastorius	5:19
Fra Mauro	9:23
Franklin	4:18
Gassendi	11:25
Geminus	3:17
Goclenius	4:18
Grimaldi	13-14:27-28
Gutenberg	5:19
Hercules	5:19
Herodotus	11:25
Hipparchus	7:21
Hommel	5:19
Humboldt	3:15
Janssen	4:18
Julius Caesar	6:20
Kepler	10:24
Landsberg	10:24
Langrenus	3:17
Letronne	11:25
Linné	6
Longomontanus	9:23

The numbers indicate the age of the Moon when features are usually best visible.

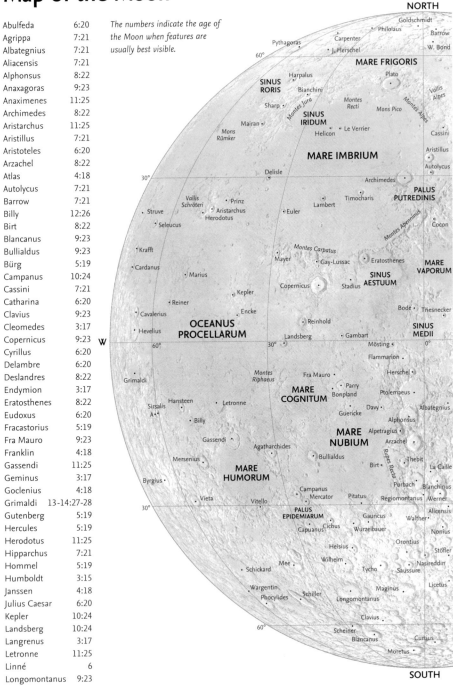

18

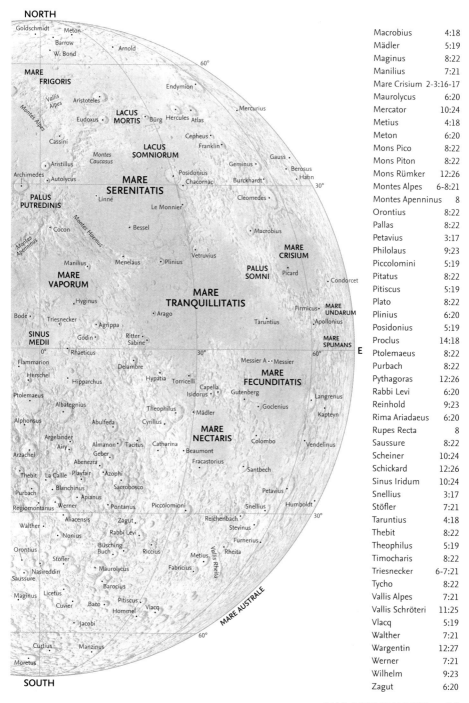

NORTH

SOUTH

Macrobius	4:18
Mädler	5:19
Maginus	8:22
Manilius	7:21
Mare Crisium	2-3:16-17
Maurolycus	6:20
Mercator	10:24
Metius	4:18
Meton	6:20
Mons Pico	8:22
Mons Piton	8:22
Mons Rümker	12:26
Montes Alpes	6-8:21
Montes Apenninus	8
Orontius	8:22
Pallas	8:22
Petavius	3:17
Philolaus	9:23
Piccolomini	5:19
Pitatus	8:22
Pitiscus	5:19
Plato	8:22
Plinius	6:20
Posidonius	5:19
Proclus	14:18
Ptolemaeus	8:22
Purbach	8:22
Pythagoras	12:26
Rabbi Levi	6:20
Reinhold	9:23
Rima Ariadaeus	6:20
Rupes Recta	8
Saussure	8:22
Scheiner	10:24
Schickard	12:26
Sinus Iridum	10:24
Snellius	3:17
Stöfler	7:21
Taruntius	4:18
Thebit	8:22
Theophilus	5:19
Timocharis	8:22
Triesnecker	6-7:21
Tycho	8:22
Vallis Alpes	7:21
Vallis Schröteri	11:25
Vlacq	5:19
Walther	7:21
Wargentin	12:27
Werner	7:21
Wilhelm	9:23
Zagut	6:20

MAP OF THE MOON **19**

Eclipses in 2020

Lunar eclipses

There are four lunar eclipses in 2020, but none are total. Technically, all are **penumbral** eclipses – and so no change in brightness is detectable with the naked eye – but in one case, the eclipse of January 10, at mid-eclipse (at 19:10 UT) it is just possible (but unlikely) that part of the Moon may darken sufficiently for the change to be detectable. But this will depend on conditions in the Earth's atmosphere and how much light it blocks.

Solar eclipses

There are just two solar eclipses in 2020. The first is an **annular** eclipse on June 21, with a path that stretches from East Africa across Arabia, India and China. Maximum eclipse occurs on the India/China border, but even here the duration is just 38 seconds.

The second eclipse, on December 14, is **total** but largely over the oceans. It starts in mid-Pacific, crosses land over Chile and Argentina (where mid-eclipse occurs, with a duration of 2 minutes 10 seconds), and ends over the Atlantic, west of Namibia.

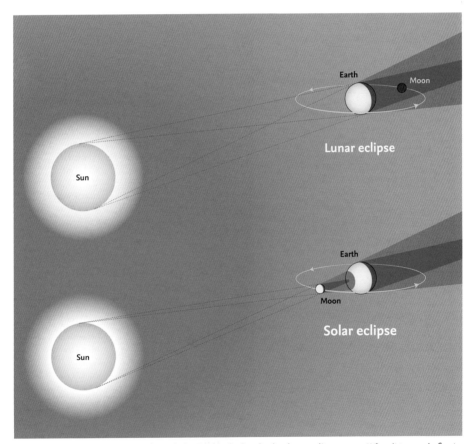

Eclipses. *When the Moon passes through the Earth's shadow (top), a lunar eclipse occurs. When it passes in front of the Sun (below) a solar eclipse occurs.*

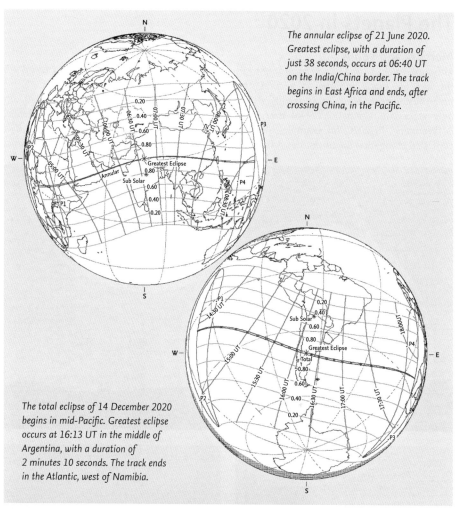

The annular eclipse of 21 June 2020. Greatest eclipse, with a duration of just 38 seconds, occurs at 06:40 UT on the India/China border. The track begins in East Africa and ends, after crossing China, in the Pacific.

The total eclipse of 14 December 2020 begins in mid-Pacific. Greatest eclipse occurs at 16:13 UT in the middle of Argentina, with a duration of 2 minutes 10 seconds. The track ends in the Atlantic, west of Namibia.

An annular eclipse of the Sun, photographed from California, towards the end of the eclipse at sunset over the Pacific.

The Planets in 2020

Mercury and Venus

Although Venus may be prominent in the sky for many months at a time, the innermost planet, Mercury, is readily visible only at certain elongations. Not every one is favourable. In 2020, for example, Mercury comes to elongation six times, but it is easily seen at four of those only. These occasions are shown on the diagrams below.

Venus has two elongations in 2020, on March 24 and August 13. By coincidence, these occur around the time of Mercury's elongations, so are also shown in the diagrams.

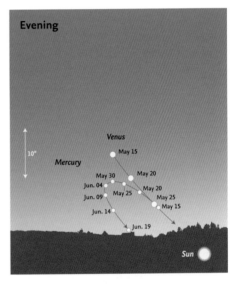

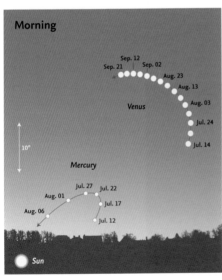

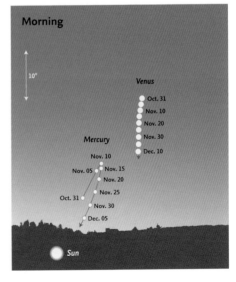

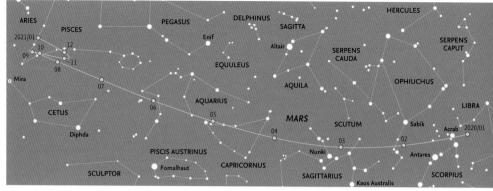

The path of Mars in 2020.

Mars

Because its orbit lies outside that of the Earth, so it takes longer to complete a full orbit, Mars does not come to opposition every year, but remains in the sky for many months at a time, moving slowly along the ecliptic. The planet's path in 2020 is shown on the large chart given here.

Oppositions occur during a period of *retrograde motion*, when the planet appears to move westwards against the pattern of distant stars. In 2020 it begins retrograde motion late in the year, on September 11, comes to opposition on October 13, and ends retrograde motion – reverting to **direct motion**

eastwards – on November 17. The period of retrograde motion and opposition is shown on the enlarged chart below (left).

Because of its eccentric orbit, which carries it at very differing distances from the Sun (and Earth), not all oppositions of Mars are equally favourable for observation. The relative positions of Mars and the Earth are shown here (below, right). It can be seen that the opposition of 2018 was very close and thus favourable for observation, and that of 2020 is also reasonably good. By comparison, opposition in 2027 will be at a far greater distance, so the planet will appear much smaller.

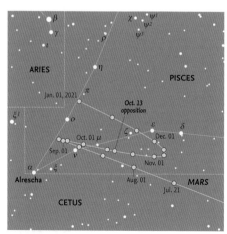

The path of Mars in the second half of 2020.
Background stars are shown down to magnitude 6.5.

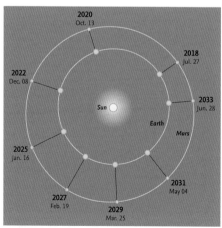

The oppositions of Mars between 2018 and 2033.

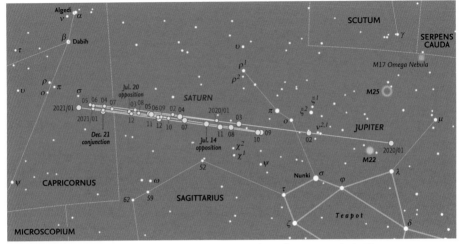

The paths of Jupiter and Saturn in 2020. Background stars are shown down to magnitude 6.5.

Jupiter and Saturn

In 2020, Jupiter and Saturn are close together in the sky in the constellations of Sagittarius and Capricornus. Jupiter comes to opposition on July 14 and Saturn on July 20. They come within 0.1° of one another on December 21.

Jupiter's four large satellites are readily visible in binoculars, although not all four are visible all the time, sometimes hidden behind the planet or invisible in front of it. Io, the closest to Jupiter, orbits the planet in just under 1.8 days, and Callisto, the farthest away, orbits in about 16.7 days. In between are Europa (c. 3.6 days) and Ganymede, the largest (c. 7.1 days). The diagram below shows the satellites' motions around the time of opposition.

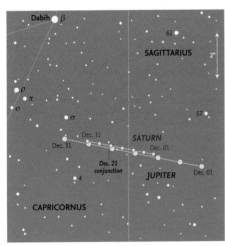

The paths of Jupiter and Saturn in December 2020.
They come to conjunction on December 21.
Background stars are shown down to magnitude 8.5.

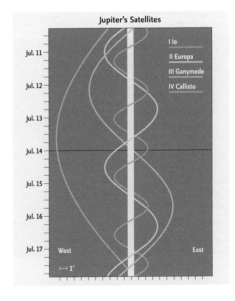

Uranus and Neptune

Uranus starts to retrograde on August 16, and comes to opposition on October 31 at mag. 5.7 in the constellation of **Aries**. Full Moon occurs that day, so it will be difficult to detect the planet. It is at the same magnitude for an extended period of the year (from August to December) and should be visible when free from interference by moonlight. It is still retrograding at the end of the year. **Neptune** begins retrograde motion on June 24 and reaches opposition at mag. 7.8 on September 11 in **Aquarius**. This is one day after Last Quarter, so there should not be too much interference from moonlight. It resumes direct motion on November 30.

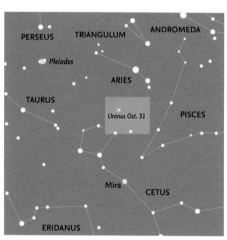

In 2020, Uranus may be found in the southern part of the constellation of Aries.

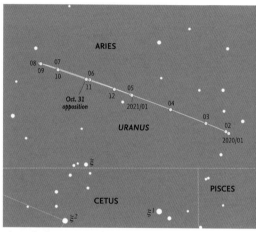

The path of Uranus in 2020. Uranus comes to opposition on October 31. All stars brighter than magnitude 7.5 are shown.

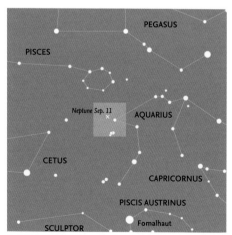

In 2020, Neptune is to be found in the easternmost part of the constellation of Aquarius.

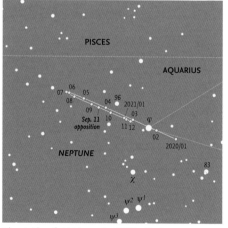

The path of Neptune in 2020. Neptune comes to opposition on September 11. All stars brighter than magnitude 8.5 are shown.

Minor Planets in 2020

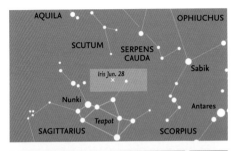

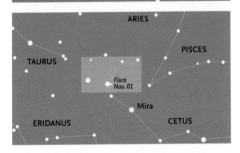

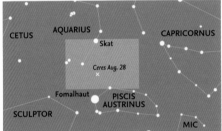

In 2020, three minor planets become bright enough to be visible at opposition. These (in date order of opposition) are: *(7) Iris* in *Sagittarius* on June 28; *(1) Ceres*, the dwarf planet, in *Aquarius* on August 28; and *(8) Flora* in *Cetus* on November 1. The locations are shown on the accompanying charts. Oppositon of (7) Iris on June 28 is at First Quarter, and for (1) Ceres on August 28, the Moon is waxing *gibbous*, so observing conditions are not ideal. With (8) Flora's opposition on November 1, Full Moon occurred the previous day, so there will be considerable interference from moonlight. It should, however, be detectable earlier than that date and also later in November.

The three maps on the left show the areas covered by the larger maps on these pages, in a lighter blue. The white cross shows the position of the minor planet at the day of its opposition.

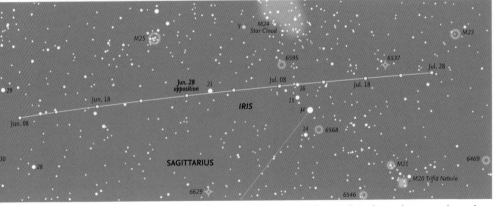

The path of the minor planet (7) Iris around its opposition on June 28 (mag. 8.8). Background stars are shown down to magnitude 9.5.

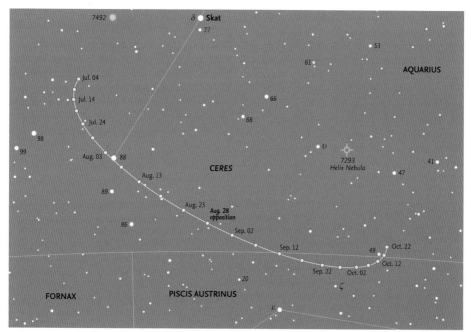

The path of the minor planet (1) Ceres around its opposition on August 28 (mag. 7.7). Background stars are shown down to magnitude 8.5.

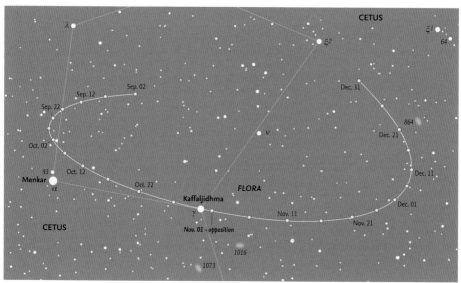

The path of the minor planet (8) Flora around its opposition on November 1 (mag. 8.0). Background stars are shown down to magnitude 9.5.

Comets in 2020

Although comets may occasionally become very striking objects in the sky, their occurrence and particularly the existence or length of any tail and their overall magnitude are notoriously difficult to predict. Naturally, it is only possible to predict the return of periodic comets (whose names have the prefix 'P'). Many comets appear unexpectedly (these have names with the prefix 'C'). Bright, readily visible comets such as C/1995 Y1 Hyakutake and C/1995 01 Hale-Bopp or C/2006 P1 McNaught (sometimes known as the Great Comet of 2007) are rare. (Comet Hale-Bopp, in particular, was visible for a record 18 months and was a prominent object in northern skies.) Comet C2006/ P1 McNaught was notable for its multiple tail structure. Most periodic comets are faint and only a very small number ever become bright enough to be readily visible with the naked eye or with binoculars.

Just one comet, **88P/PanSTARRS**, may become brighter than magnitude 8.2 at the beginning of May 2020 and will be visible from the northern hemisphere.

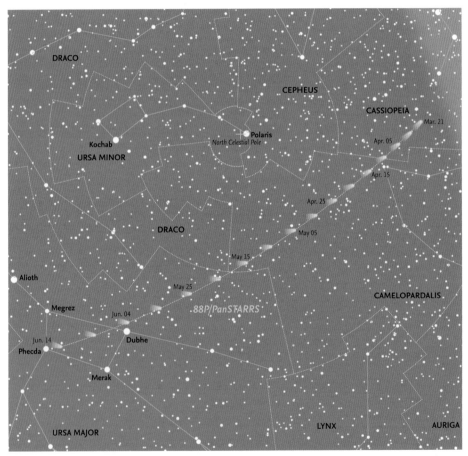

Comet 88P/PanSTARRS will not become brighter than magnitude 8.2, in the beginning of May. The chart shows its path from March 21 to June 14. Background stars are shown down to magnitude 7.5.

Comet C/2014 Q2 Lovejoy, which reached naked-eye visibility, photographed on 20 December 2014, when in Columba, by Damian Peach.

Comet C/2018 Y1 Iwamoto, photographed in February 2018, when near the 'Flaming Star' nebulosity in Auriga (Photographer: Denis Buczynski).

Introduction to the Month-by-Month Guide

The monthly charts

The pages devoted to each month contain a pair of charts showing the appearance of the night sky, looking north and looking south. The charts (as with all the charts in this book) are drawn for the latitude of 50°N, so observers farther north will see slightly more of the sky on the northern horizon, and slightly less on the southern. These areas are, of course, those most likely to be affected by poor observing conditions caused by haze, mist or smoke. In addition, stars close to the horizon are always dimmed by atmospheric absorption, so sometimes the faintest stars marked on the charts may not be visible.

The three times shown for each chart require a little explanation. The charts are drawn to show the appearance at 23:00 GMT for the 1st of each month. The same appearance will apply an hour earlier (22:00 GMT) on the 15th, and yet another hour earlier (21:00 GMT) at the end of the month (shown as the 1st of the following month). GMT is identical to the Universal Time (UT) used by astronomers around the world. In Europe, Summer Time is introduced in March, so the March charts apply to 23:00 GMT on March 1, 22:00 GMT on March 15, but 22:00 BST (British Summer Time) on April 1. The change back from Summer Time (in Europe) occurs in October, so the charts for that month apply to 00:00 BST for October 1, 23:00 BST for October 15, and 21:00 GMT for November 1.

The charts may be used for earlier or later times during the night. To observe two hours earlier, use the charts for the preceding month; for two hours later, the charts for the next month.

Meteors

Details of specific meteor showers are given in the months when they come to maximum, regardless of whether they begin or end in other months. Note that not all the respective radiants are marked on the charts for that

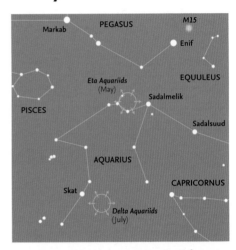

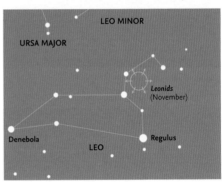

particular month, because the radiants may be below the horizon, or lie in constellations that are not readily visible during the month of maximum. For this reason, special charts for the Eta and Delta Aquariids (May and July, respectively) and the Leonids (November) are given here. As explained earlier, however, meteors from such showers may still be seen, because the most effective region for seeing meteors is some 40–45° away from the radiant and that area of sky may well be above the horizon. A table of the best meteor showers visible during the year is also given here. The rates given are based on the properties of the meteor streams, and are those that an experienced observer might see under ideal

Shower	Dates of activity 2020	Date of maximum 2020	Possible hourly rate
Quadrantids	December 28 to January 12	January 4	120
April Lyrids	April 13–29	April 22	18
Eta Aquariids	April 18 to May 27	May 5	55
Alpha Capricornids	July 2 to August 14	July 29	5
Delta Aquariids	July 13 to August 24	July 31	< 20
Perseids	July 16 to August 23	August 12	100
Alpha Aurigids	August 28 to September 5	August 31	10
Southern Taurids	September 10 to November 20	October 10	< 5
Orionids	October 1 to November 6	October 21	25
Draconids	October 7–11	October 10	10
Northern Taurids	October 20 to December 10	November 13	< 5
Leonids	November 5–29	November 17	< 15
Geminids	December 3–16	December 13	100+
Ursids	December 17–26	December 22	< 10

conditions. Generally the observed rates will be far less.

Meteors that are brighter than magnitude -4 (approximately the maximum magnitude reached by Venus) are known as *fireballs* or *bolides*. Examples are shown on pages 37 and 77. Fireballs sometimes cause sonic booms that may be heard some time after the meteor is seen.

The photographs

As an aid to identification – especially as some people find it difficult to relate charts to the actual stars they see in the sky – one or more photographs of constellations visible in certain specific months are included. It should be noted, however, that because of the limitations of the photographic and printing processes, and the differences between the sensitivity of different individuals to faint starlight (especially in their ability to detect different colours), and the degree to which they have become adapted to the dark, the apparent brightness of stars in the photographs will not necessarily precisely match that seen by any one observer.

The Moon calendar

The Moon calendar is largely self-explanatory. It shows the phase of the Moon for every day of the month, with the exact times (in Universal Time) of New Moon, First Quarter, Full Moon and Last Quarter. Because the

times are calculated from the Moon's actual orbital parameters, some of the times shown will, naturally, fall during daylight, but any difference is too small to affect the appearance of the Moon on that date. Also shown is the *age* of the Moon (the day in the *lunation*), beginning at New Moon, which may be used to determine the best time for observation of specific lunar features.

The Moon

The section on the Moon includes details of any lunar or solar eclipses that may occur during the month (visible from anywhere on Earth). Similar information is given about any important occultations. Mainly, however, this section summarizes when the Moon passes close to planets or the five prominent stars close to the ecliptic. The dates when the Moon is closest to the Earth (at *perigee*) and farthest from it (at *apogee*) are shown in the monthly calendars, and only mentioned here when they are particularly significant, such as the nearest and farthest during the year.

The planets and minor planets

Brief details are given of the location, movement and brightness of the planets from Mercury to Saturn throughout the month. None of the planets can, of course, be seen when they are close to the Sun, so such periods are generally noted. All of the planets

A fireball (with flares approximately as bright as the Full Moon), photographed against a weak auroral display by D. Buczynski from Tarbat Ness in Scotland on 22 January, 2017.

may sometimes lie on the opposite side of the Sun to the Earth (at superior conjunction), but in the case of the inferior planets, Mercury and Venus, they may also pass between the Earth and the Sun (at inferior conjunction) and are invisible for a period of time, the length of which varies from conjunction to conjunction. Those two planets are normally easiest to see around either eastern or western elongation, in the evening or morning sky, respectively. Not every elongation is favourable, so although every elongation is listed, only those where observing conditions are favourable are shown in the individual diagrams of events.

The dates at which the superior planets reverse their motion (from direct motion to *retrograde*, and retrograde to direct) and of opposition (when a planet generally reaches its maximum brightness) are given. Some planets, especially distant Saturn, may spend most or all of the year in a single constellation. Jupiter and Saturn are normally easiest to see around opposition, which occurs every year. Mars, by contrast, moves relatively rapidly against the background stars and in some years never comes to opposition.

Uranus is normally magnitude 5.7–5.9, and thus at the limit of naked-eye visibility under exceptionally dark skies, but bright enough to be readily visible in binoculars. Because its orbital period is so long (over 84 years), Uranus moves only slowly along the ecliptic, and often remains within a single constellation for a whole year. The chart on page 25 shows its position during 2020.

Similar considerations apply to Neptune, although it is always fainter (generally magnitude 7.8–8.0), still visible in most binoculars. It takes about 164.8 years to complete one orbit of the Sun. As with Uranus, it frequently spends a complete year in one constellation. Its chart is also on page 25.

In any year, few minor planets ever become bright enough to be detectable in binoculars. Just one, (4) Vesta, on rare occasions brightens sufficiently for it to be visible to the naked eye. Our limit for visibility is magnitude 9.0 and details and charts are given for those objects that exceed that magnitude during the year, normally around opposition. To assist in recognition of a planet or minor planet as it moves against the background stars, the latter are shown to a fainter magnitude than the object at opposition. Minor-planet charts for 2020 are on pages 26 and 27.

The ecliptic charts

Although the ecliptic charts are primarily designed to show the positions and motions of the major planets, they also show the motion of the Sun during the month. The light-tinted area shows the area of the sky that is invisible during daylight, but the darker area gives an indication of which constellations are likely to be visible at some time of the night. The closer a planet is to the border between dark and light, the more difficult it will be to see in the twilight.

The monthly calendar

For each month, a calendar shows details of significant events, including when planets are close to one another in the sky, close to the Moon, or close to any one of five bright stars that are spaced along the ecliptic. The times shown are given in Universal Time (UT), always used by astronomers throughout the year, and which is identical to Greenwich Mean Time (GMT). So during the summer months, they do not show Summer Time, which will always be one hour later than the time shown.

The diagrams of interesting events

Each month, a number of diagrams show the appearance of the sky when certain events take place. However, the exact positions of celestial objects and their separations greatly depend on the observer's position on Earth. When the Moon is one of the objects involved, because it is relatively close to Earth, there may be very significant changes from one location to another. Close approaches between planets or between a planet and a star are less affected by changes of location, which may thus be ignored.

The diagrams showing the appearance of the sky are drawn for the latitude of London, so will be approximately correct for most of Britain and Europe. However, for an observer farther north (say, Edinburgh), a planet or star listed as being north of the Moon will appear even farther north, whereas one south of the Moon will appear closer to it – or may even be hidden (occulted) by it. For an observer farther south than London, there will be corresponding changes in the opposite direction: for a star or planet south of the Moon the separation will increase, and for one north of the Moon the separation will decrease. This is particularly important when occultations occur, which may be visible from one location, but not another. However, there are no major occultations visible from Britain in 2020.

Ideally, details should be calculated for each individual observer, but this is obviously impractical. In fact, positions and separations are actually calculated for a theoretical observer located at the centre of the Earth.

So the details given regarding the positions of the various bodies should be used as a guide to their location. A similar situation arises with the times that are shown. These are calculated according to certain technical criteria, which need not concern us here. However, they do not necessarily indicate the exact time when two bodies are closest together. Similarly, dates and times are given, even if they fall in daylight, when the objects are likely to be completely invisible. However, such times do give an indication that the objects concerned will be in the same general area of the sky during both the preceding, and the following nights.

Key to the symbols used on the monthy star maps.

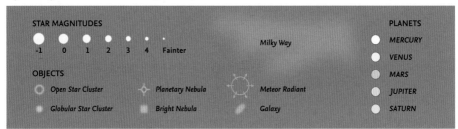

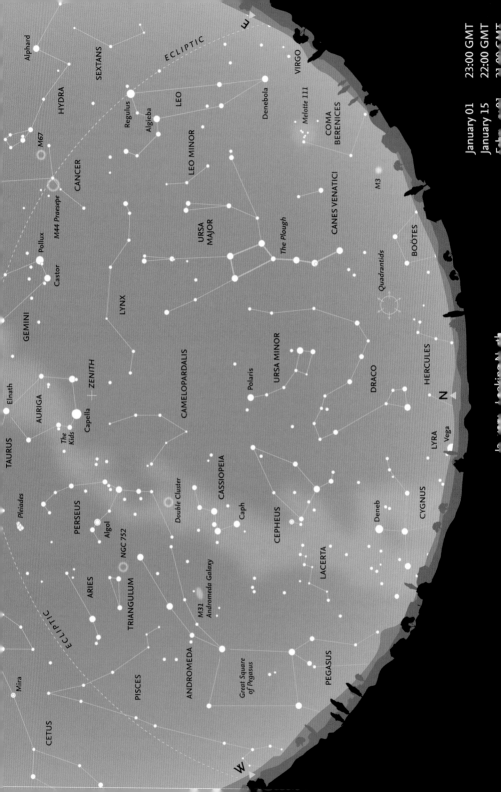

January 01   23:00 GMT
January 15   22:00 GMT

January – Looking North

The constellation of Orion dominates the sky during this period of the year, and is a useful starting point for recognizing other constellations in the southern sky. Here, orange Betelgeuse, blue-white Rigel and the pinkish Orion Nebula are prominent. Orion can be found in the southern part of the sky (see next page).

Most of the important circumpolar constellations are easy to see in the northern sky at this time of year. **Ursa Major** stands more-or-less vertically above the horizon in the northeast, with the zodiacal constellation of **Leo** rising in the east. To the north, the stars of **Ursa Minor** lie below **Polaris** (the Pole Star). The head of **Draco** is low on the northern horizon, but may be difficult to see unless observing conditions are good. Both **Cepheus** and **Cassiopeia** are readily visible in the northwest, and even the faint constellation of **Camelopardalis** is high enough in the sky for it to be easily visible.

Near the zenith is the constellation of **Auriga** (the Charioteer), with brilliant **Capella** (α Aurigae) directly overhead. Slightly to the west of Capella lies a small triangle of fainter stars, known as 'The Kids'. (Ancient mythological representations of Auriga show him carrying two young goats.) Together with the northernmost bright star in Taurus, **Elnath** (β Tauri), the body of Auriga forms a large pentagon on the sky, with The Kids lying on the western side. Farther down towards the west are the constellations of **Perseus** and **Andromeda**, and the Great Square of **Pegasus** is approaching the horizon.

Meteors

One of the strongest and most consistent meteor showers of the year occurs in January: the **Quadrantids**, which are visible January 1–12, with a maximum on January 3–4. They are brilliant, bluish and yellowish-white meteors and fireballs (page 37), which at maximum may even reach a rate of 120 meteors per hour. At maximum, the Moon is waxing gibbous, just past First Quarter, so there will be some interference from moonlight. The parent object is minor planet 2003 EH$_1$.

The shower is named after the former constellation **Quadrans Muralis** (the Mural Quadrant), an early form of astronomical instrument. The Quadrantid meteor radiant is now within the northernmost part of **Boötes**, roughly halfway between θ Boötis and τ Herculis.

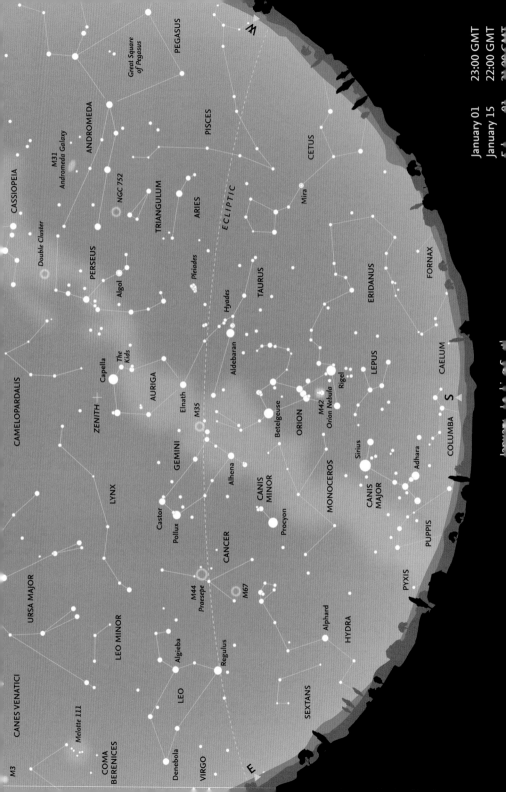

W

PEGASUS
Great Square
of Pegasus

ANDROMEDA
CASSIOPEIA
M31
Andromeda Galaxy
Double Cluster
NGC 752
PERSEUS
Algol
CAMELOPARDALIS
TRIANGULUM
ARIES
PISCES
ECLIPTIC
Pleiades
Mira
CETUS
The
Kids
Hyades
Capella
TAURUS
Aldebaran
ZENITH
AURIGA
Elnath
M35
FORNAX
ERIDANUS
LYNX
GEMINI
Betelgeuse
ORION
M42
Orion Nebula
Rigel
LEPUS
CAELUM
Castor
Alhena
MONOCEROS
Pollux
CANIS
MINOR
Sirius
COLUMBA
Procyon
CANIS
MAJOR
Adhara
S
CANCER
M44
Praesepe
M67
HYDRA
PUPPIS
PYXIS
LEO MINOR
URSA MAJOR
Regulus
Algieba
LEO
Alphard
SEXTANS
CANES VENATICI
Melotte 111
Denebola
VIRGO
COMA
BERENICES
M3

E

January – Looking South

At this time of year the southern sky is dominated by **Orion**. This is the most prominent constellation during the winter months, when it is visible at some time during the night. (A photograph of Orion appears on page 35.) It has a highly distinctive shape, with a line of three stars that form the 'Belt'. To most observers, the bright star at the northeastern corner of the constellation, **Betelgeuse** (α Orionis), shows a reddish tinge, in contrast to the brilliant bluish-white colour of the bright star at the southwestern corner, **Rigel** (β Orionis). The three stars of the belt lie directly south of the celestial equator. A vertical line of three 'stars' forms the 'Sword' that hangs to the south of the Belt. With good viewing conditions, the central 'star' appears as a hazy spot, even to the naked eye. This is actually the **Orion Nebula**. Binoculars will reveal the four stars of the '**Trapezium**', which illuminate the nebula.

The line of Orion's Belt points up to the northwest towards **Taurus** (the Bull) and below orange-tinted **Aldebaran** (α Tauri). Close to Aldebaran, there is a conspicuous 'V' of stars, pointing down to the southwest, called the **Hyades** cluster. (Despite appearances, Aldebaran is not part of the cluster.) Farther along, the same line from Orion

A late Quadrantid fireball, photographed from Portmahomack, Ross-shire, Scotland on January 15, 2018 at 23:44 UT.

passes below a bright cluster of stars, the **Pleiades**, or Seven Sisters. Even the smallest pair of binoculars reveals this cluster to be a beautiful group of bluish-white stars. The two most conspicuous of the other stars in Taurus lie directly above Orion, and form an elongated triangle with Aldebaran. The northernmost, β Tauri, was once considered to be part of the constellation of Auriga.

The Moon's phases for January

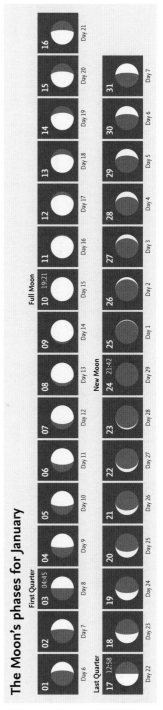

First Quarter			
01	02	**03** 04:45	04
Day 5	Day 6	Day 7	Day 8

05	06	07	08
Day 9	Day 10	Day 11	Day 12

Full Moon

09	**10** 19:21	11	12
Day 13	Day 14	Day 15	Day 16

13	14	15	16
Day 17	Day 18	Day 19	Day 20

Last Quarter

17 12:58	18	19	20
Day 21	Day 22	Day 23	Day 24

21	22	23	**24** 21:42
Day 25	Day 26	Day 27	Day 28

New Moon

25	26	27	28
Day 29	Day 1	Day 2	Day 3

29	30	31	
Day 4	Day 5	Day 6	Day 7

January – Moon and Planets

The Earth

The Earth reaches *perihelion* (the closest point in its orbit to the Sun) on January 5 at 07:48 Universal Time, when its distance is 0.9832 AU (147,091,074 km).

The Moon

On January 7, the Moon is 5° north of *Aldebaran*. At Full Moon on January 10 there is a penumbral lunar eclipse, the end of which is theoretically visible from Britain. A small part of the Moon may pass within the umbra and thus darken at mid-eclipse (19:10 UT). The Moon passes *Regulus* three days later and Mars again (but this time as a waning crescent and morning object) on January 20. It is due south of Venus on January 28 as a waxing crescent in the evening sky.

The planets

Mercury is close to the Sun and invisible this month. (It is at superior conjunction on the far side of the Sun on January 10.) At the end of the month it begins to become visible in the evening sky, before eastern elongation in February. *Venus* is in the evening sky, moving from *Capricornus* into *Aquarius* and away from the Sun, brightening slightly from mag. –4.0 to –4.1. It is 4.1° north of the Moon on January 28. *Mars* is a morning object in *Libra* and then moving into *Scorpius*, being due north of *Antares* (the 'Rival of Mars') on January 17. *Jupiter* and *Saturn* (both in *Sagittarius*) are too close to the Sun to be visible. (Saturn is at superior conjunction on January 13.) *Uranus* is in Aries (where it is for the whole year) at mag. 5.7–5.8. *Neptune* is in *Aquarius* at mag. 7.9.

The path of the Sun and the planets along the ecliptic in January.

2	01:30	Moon at apogee (404,580 km)
3	04:45	First Quarter
4	08:31	Quadrantid shower maximum
5	07:48	Earth at perihelion (147,091,074 km = 0.9832 AU)
7	21:39	Aldebaran 5.0°S of Moon
0	15:19	Mercury at superior conjunction
0	19:10	Penumbral lunar eclipse
0	19:21	Full Moon
0	02:54	Pollux 5.3°N of Moon
2		Quadrantid meteor shower ends
3	12:03	Regulus 3.8°S of Moon
3	14:23	Saturn at conjunction with Sun
3	20:20	Moon at perigee (366,000 km)
7	03:25	Spica 7.7°S of Moon
7	04:00 ★	Antares 4.8°S of Mars
7	12:58	Last Quarter
0	14:39	Antares 7.2°S of Moon
0	19:12	Mars 2.3°S of Moon
3	02:41	Jupiter 0.4°N of Moon
4	21:42	New Moon
4	01:39	Saturn 1.5°N of Moon
5	18:13	Mercury 1.3°N of Moon
8	07:28	Venus 4.1°N of Moon
9	21:27	Moon at apogee (405,392 km)

These objects are close together for an extended

Morning 7:00

January 13–14 • The Moon passes between Regulus and Algieba in the western sky.

Evening 23:00

January 7 • The Moon with Aldebaran and the Pleiades. Betelgeuse is farther south.

Evening 18:00

January 27–28 • The Moon with Venus in the evening sky.

Morning 7:00

January 20–21 • The Moon passes Acrab (β Sco), Antares, Mars and Sabik (η Oph) in the Morning sky. Mars and Sun...

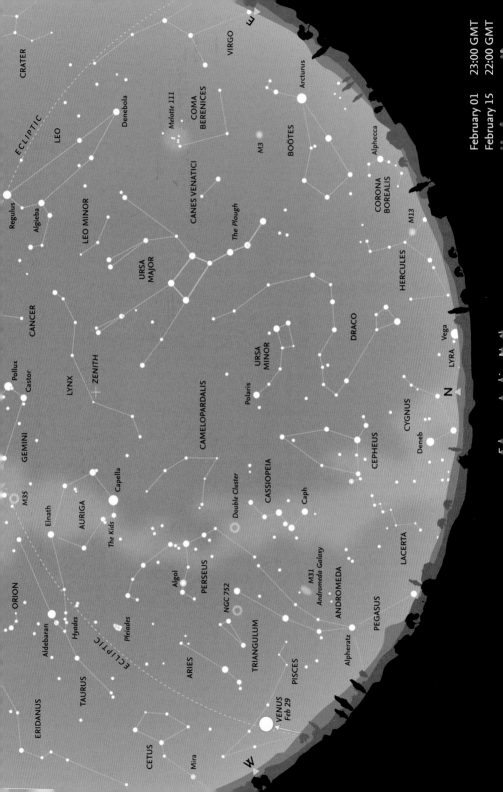

CRATER

VIRGO

ECLIPTIC

LEO

Denebola

Arcturus

Melotte 111

COMA
BERENICES

M3

BOÖTES

Regulus

Algieba

CANES VENATICI

Alphecca

The Plough

CORONA
BOREALIS

LEO MINOR

URSA
MAJOR

M13

HERCULES

CANCER

DRACO

LYNX

ZENITH

URSA
MINOR

Pollux

Castor

CAMELOPARDALIS

Vega

LYRA

Polaris

N

GEMINI

CYGNUS

CEPHEUS

M35

Capella

Deneb

Elnath

AURIGA

Double Cluster

CASSIOPEIA

Caph

The Kids

LACERTA

Algol

PERSEUS

Andromeda Galaxy

M31

ANDROMEDA

ORION

NGC 752

PEGASUS

Aldebaran

Hyades

TRIANGULUM

Alpheratz

TAURUS

Pleiades

ECLIPTIC

ARIES

PISCES

ERIDANUS

VENUS
Feb 29

CETUS

Mira

W

February 01   23:00 GMT
February 15   22:00 GMT

February – Looking North

The months of January and February are probably the best time for seeing the section of the Milky Way that runs in the northern and western sky from **Cygnus**, low on the northern horizon, through **Cassiopeia**, **Perseus** and **Auriga** and then down through **Gemini** and **Orion**. Although not as readily visible as the denser star clouds of the summer Milky Way, on a clear night so many stars may be seen that even a distinctive constellation such as Cassiopeia is not immediately obvious.

The head of **Draco** is now higher in the sky and easier to recognize. **Deneb** (α Cygni), the brightest star in **Cygnus**, may just be visible almost due north at midnight, early in the month, if the sky is very clear and the horizon clear of obstacles. **Vega** (α Lyrae) in **Lyra** is so low that it is difficult to see, but may become visible later in the night. The constellation of **Boötes** – sometimes described as shaped like a kite, an ice-cream cone, or the letter 'P' – with orange-tinted **Arcturus** (α Boötis), is beginning to clear the eastern horizon. Arcturus, at magnitude -0.05, is the brightest star in the northern hemisphere. The inconspicuous constellation of **Coma Berenices** is now well above the horizon in the east. The concentration of faint stars at the northwestern corner somewhat resembles a tiny, detached portion of the Milky Way. This is Melotte 111, an open star cluster (which is sometimes called the Coma Cluster, but must not be confused with the important Coma Cluster of galaxies, Abell 1656, mentioned on page 55).

On the other side of the sky, in the northwest, most of the constellation of **Andromeda** is still easily seen, although **Alpheratz** (α Andromedae), the star that forms the northeastern corner of the Great Square of Pegasus – even though it is actually part of Andromeda – is becoming close to the horizon and more difficult to detect. High overhead, at the zenith, try to make out the very faint constellation of

Lynx. It was introduced in 1687 by the famous astronomer Johannes Hevelius to fill the largely blank area between **Auriga**, **Gemini** and **Ursa Major**, and is reputed to be so named because one needed the eyes of a lynx to detect it.

A very large, and frequently ignored, open star cluster, Melotte 111, also known as the Coma Cluster, is readily visible in the eastern sky during February.

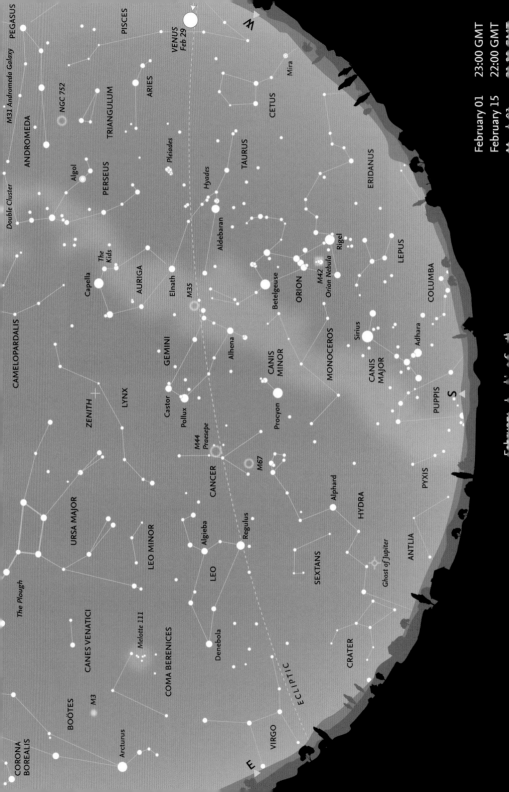

February – Looking South

Apart from **Orion**, the most prominent constellation visible this month is **Gemini**, with its two lines of stars running southwest towards Orion. Many people have difficulty in remembering which is which of the two stars **Castor** and **Pollux**. Think of them in alphabetical order: Castor (α Geminorum), the fainter star, is closer to the North Celestial Pole. Pollux (β Geminorum) is the brighter of the two, but is farther away from the Pole. Castor is remarkable because it is actually a multiple system, consisting of no less than six individual stars.

Using Orion's belt as a guide, it points down to the southeast towards **Sirius**, the brightest star in the sky (at magnitude -1.4) in the constellation of **Canis Major**, the whole of which is now clear of the southern horizon. Forming an equilateral triangle with **Betelgeuse** in Orion and Sirius in Canis Major is **Procyon**, the brightest star in the small constellation of **Canis Minor**. Between Canis Major and Canis Minor is the faint constellation of **Monoceros**, which actually straddles the Milky Way, which, although faint, has many clusters in this area. Directly east of Procyon is the highly distinctive asterism of six stars that form the 'head' of **Hydra**, the largest

The constellation of Gemini. The two brightest stars are Castor and Pollux, visible on the left-hand side of the photograph.

of all 88 constellations, and which trails such a long way across the sky that it is only in mid-March around midnight that the whole constellation becomes visible.

The Moon's phases for February

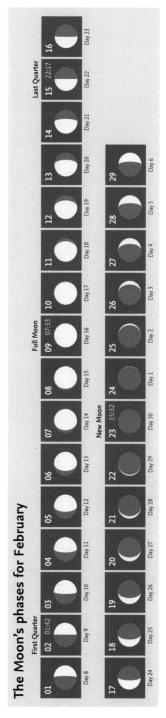

First Quarter

| 01 | 02 01:42 | 03 | 04 | 05 | 06 | 07 | 08 |
| Day 8 | Day 9 | Day 10 | Day 11 | Day 12 | Day 13 | Day 14 | Day 15 |

New Moon

Full Moon

| 09 07:33 | 10 | 11 | 12 | 13 | 14 | 15 22:17 | 16 |
| Day 16 | Day 17 | Day 18 | Day 19 | Day 20 | Day 21 | Day 22 | Day 23 |

Last Quarter

| 17 | 18 | 19 | 20 | 21 | 22 | 23 15:32 | 24 |
| Day 24 | Day 25 | Day 26 | Day 27 | Day 28 | Day 29 | Day 30 | Day 1 |

| 25 | 26 | 27 | 28 | 29 |
| Day 2 | Day 3 | Day 4 | Day 5 | Day 6 |

February – Moon and Planets

The Moon

On February 4, the Moon is 3.1°N of *Aldebaran*. Five days later (on February 9) it passes 3.8°N of *Regulus*, between it and *Algieba* (γ Leonis). On February 13, in the morning sky, it passes 7.6°N of *Spica* in *Virgo* and three days later (February 16) a similar distance (7.1°) north of *Antares* in *Scorpius*. On February 18–20 it passes, in succession, *Mars*, *Jupiter* and *Saturn* (in *Sagittarius*). Later in the month (on February 27–28) it passes below *Venus* in the evening sky.

The planets

Mercury comes to a favourable eastern elongation on February 10 at mag. -0.6 (diagram page 22). *Venus* is brilliant in the evening sky, above Mercury, at mag. -4.1 to -4.3. *Mars*, initially in *Scorpius*, moves into *Sagittarius*, brightening slightly from mag. 1.4 to 1.1 over the month and glimpsed in the eastern sky early in the month. *Jupiter* (mag. -1.9 to -2.0), and *Saturn* (mag. 0.6–0.7) are in *Sagittarius*, too close to the Sun to be visible. *Uranus* (mag. 5.8) is in Aries and *Neptune* (mag. 8.0) remains in *Aquarius* in the daytime sky.

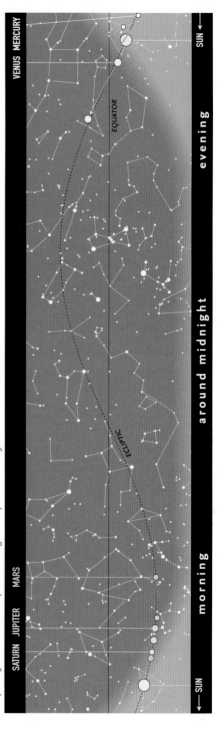

The path of the Sun and the planets along the ecliptic in February.

Calendar for February

02	01:42	First Quarter
04	07:24	Aldebaran 3.1°S of Moon
07	13:26	Moon 5.3°S of Pollux
09	07:33	Full Moon
09	21:41	Regulus 3.8°S of Moon
10	13:59	Mercury at greatest elongation (18.2°E, mag. -1.4)
10	20:30	Moon at perigee (360,500 km)
13	10:14	Spica 7.6°S of Moon
15	22:17	Last Quarter
16	20:05	Antares 7.1°S of Moon
18	13:17	Mars 0.8°S of Moon
19	19:36	Jupiter 0.9°N of Moon
20	13:39	Saturn 1.8°N of Moon
23	15:32	New Moon
23	18:43	Mercury 8.6°N of Moon
26	01:37	Mercury at inferior conjunction
26	11:35	Moon at apogee (406,300 km)
27	11:52	Venus 6.3°N of Moon

February 9 • The Moon is between Regulus and Algieba.

February 13 • The Moon with Spica in the southwest in the early morning sky.

February 18–20 • The Moon passes Mars, Nunki, Jupiter and Saturn low in the southeast about half an hour or so before sunrise.

February 27–28 • The Moon with Venus in the evening sky.

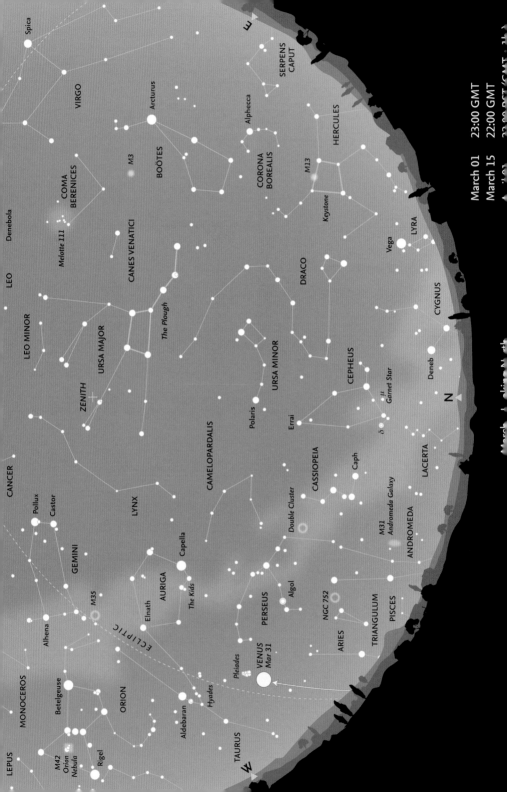

March – Looking North

In March, the Sun crosses the celestial equator on Friday, March 20, at the vernal equinox, when day and night are of almost exactly equal length, and the northern season of spring is considered to have begun. (The hours of daylight and darkness change most rapidly around the equinoxes, in March and September.) It is also in March that Summer Time begins in Europe (on Sunday, March 29) so the charts show the appearance at 23:00 GMT for March 1 and 22:00 BST for April 1. (In North America, Daylight Saving Time is introduced three weeks earlier, on Sunday, March 8.)

Early in the month, the constellation of **Cepheus** lies almost due north, with the distinctive 'W' of **Cassiopeia** to its west. Cepheus lies across the border of the Milky Way and is often described as like the gable-end of a house or a church tower and steeple. Despite the large number of stars revealed at the base of the constellation by binoculars, one star stands out because of its deep red colour. This is Mu (μ) Cephei, also known as the Garnet Star, because of its striking colour. It is a truly gigantic star, a red supergiant, and one of the largest stars known. It is about 1,400 times the diameter of the Sun, and if placed in the Solar System would extend beyond the orbit of Saturn. (Betelgeuse, in Orion, is also a red supergiant, but it is 'only' about 500 times the diameter of the Sun.)

Another famous, and very important star in Cepheus is δ **Cephei**, which is the prototype for the class of variable stars known as Cepheids. These giant stars show a regular variation in their luminosity, and there is a direct relationship between the period of the changes in magnitude and the stars' actual luminosity. From a knowledge of the period of any Cepheid, its actual luminosity – known as its absolute magnitude – may be derived. A comparison of its apparent magnitude on the sky and its absolute magnitude enables the star's exact distance to be

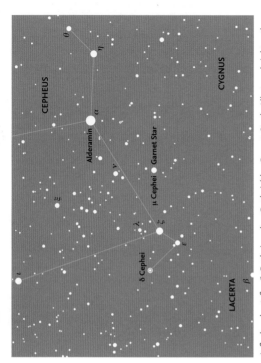

A finder chart for δ Cephei and μ Cephei (the Garnet Star). All stars brighter than magnitude 7.5 are shown.

determined. Once the distances to the first Cepheid variables had been established, examples in more distant galaxies provided information about the scale of the universe. Cepheid variables are the first 'rung' in the cosmic distance ladder. Both important stars are shown on the accompanying chart.

Below Cepheus to the east (to the right), it may be possible to catch a glimpse of **Deneb** (α Cygni), just above the horizon. Slightly farther round towards the northeast, **Vega** (α Lyrae) is marginally higher in the sky. From southern Britain, Deneb is just far enough north to be circumpolar (although difficult to see in January and February because it is so low on the northern horizon). Vega, by contrast, farther south, is completely hidden during the depths of winter.

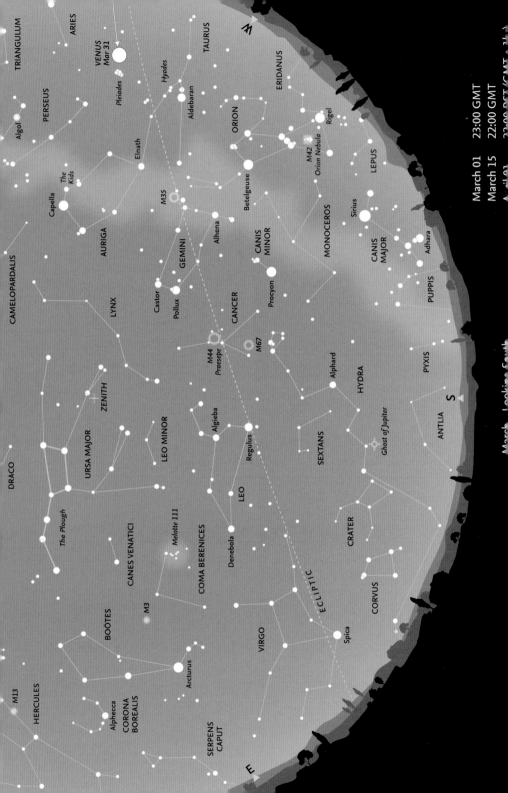

March 01   23:00 GMT
March 15   22:00 GMT

March – Looking South

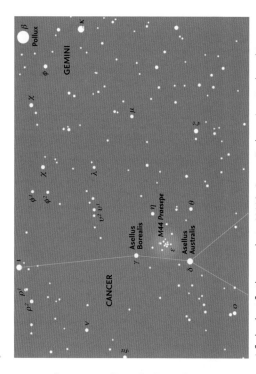

Due south at 22:00 at the beginning of the month, lying between the constellations of **Gemini** in the west and **Leo** in the east, and fairly high in the sky above the head of **Hydra**, is the faint, and rather undistinguished zodiacal constellation of **Cancer**. Rather like the triskelion, the symbol for the Isle of Man, it has three 'legs' radiating from the centre, where there is an open cluster, M44 or **Praesepe** ('the Manger' but also known as 'the Beehive'). On a clear night this cluster, known since antiquity, is just a hazy spot to the naked eye, but appears in binoculars as a group of dozens of individual stars.

Also prominent in March is the constellation of **Leo**, with the 'backward question mark' (or 'Sickle') of bright stars forming the head of the mythological lion. **Regulus** (α Leonis) – the 'dot' of the 'question mark' or the handle of the sickle and the brightest star in Leo – lies very close to the ecliptic and is one of the few first-magnitude stars that may be occulted by the Moon. However, there are no occultations of Regulus in 2020, nor of any of the other four bright stars near the ecliptic.

A finder chart for the open cluster M44 in Cancer. To the ancient Greeks and Romans, the two stars Asellus Borealis and Asellus Australis represented two donkeys feeding from Praesepe ('the Manger'). All stars brighter than magnitude 7.5 are shown.

The Moon's phases for March

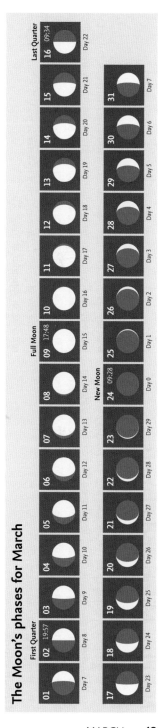

March – Moon and Planets

The Moon

On March 1 the Moon passes the *Pleiades* and on March 2 it is 3.3° north of *Aldebaran*. It is north of *Regulus* on March 8, of *Spica* on March 11, and *Antares* on March 15. When it comes to apogee on March 24, it is the most distant of the year, at 406,700 km. The Moon passes south of *Venus* and the *Pleiades* in the evening sky on March 28 and north of *Aldebaran* on March 29.

Occultations

None of the five bright stars close to the ecliptic (*Aldebaran*, *Antares*, *Pollux*, *Regulus* and *Spica*) are occulted by the Moon in 2020. There is one occultation of *Mars* on September 6, which is visible from South America only.

The planets

Mercury reaches greatest western elongation on March 24 but is too low to be readily visible. *Venus* is brilliant in the evening sky and reaches greatest eastern elongation, at mag. -4.5, on March 24 (diagram page 22). *Mars* is low in the morning sky at mag. 1.1–0.8, together with *Jupiter* (mag. -2.0 to -2.1) and *Saturn* (mag. 0.7). Saturn is 0.9° south of Mars on March 31. *Uranus* is still in *Aries* at mag. 5.8, and *Neptune* remains in *Aquarius* at mag. 8.0.

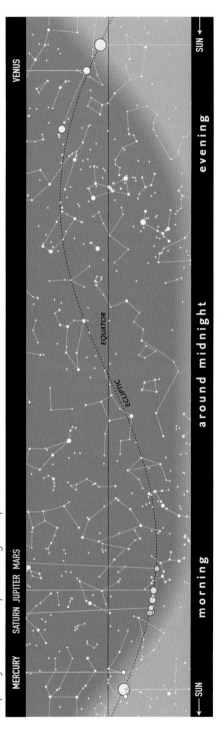

The path of the Sun and the planets along the ecliptic in March.

Calendar for March

Date	Time	Event
02	15:48	Aldebaran 3.3°S of Moon
02	19:57	First Quarter
05	23:54	Moon 5.1°S of Pollux
08		Daylight Saving Time begins (North America)
08	08:49	Regulus 3.8°S of Moon
08	12:09	Neptune at conjunction with the Sun
09	17:48	Full Moon
10	06:33	Moon at perigee (357,100 km)
11	19:42	Spica 7.4°S of Moon
15	02:44	Antares 6.8°S of Moon
16	09:34	Last Quarter
18	08:18	Mars 0.7°N of Moon
18	10:18	Jupiter 1.5°N of Moon
18	23:56	Saturn 2.1°N of Moon
20	03:50	Northern spring equinox
20	06:00 *	Mars 0.7°S of Jupiter
21	17:48	Mercury 3.6°N of Moon
24	01:59	Mercury at greatest elongation (27.8°W, mag. 0.2)
24	09:28	New Moon
24	15:23	Moon at apogee (406,700 km, greatest of year)
24	21:59	Venus at greatest elongation (46.1°E, mag. –4.5)
28	10:39	Venus 6.8°N of Moon
29		Summer Time begins (Europe)
29	22:25	Aldebaran 3.6°S of Moon
31	11:00 *	Saturn 0.9°S of Mars

These objects are close together for an extended period around this time.

Evening 20:00

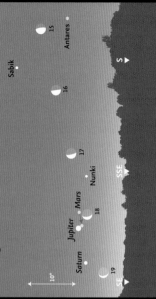

March 1–2 • The Moon passes between Aldebaran and the Pleiades.

Evening 21:00

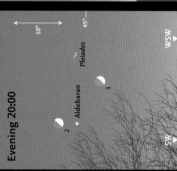

March 28–29 • The Moon with Venus, the Pleiades and Aldebaran in the western evening sky.

Morning 5:30

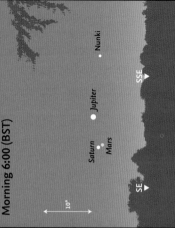

March 15–19 • The Moon passes Antares, Nunki, Mars, Jupiter and Saturn in the early morning.

Morning 6:00 (BST)

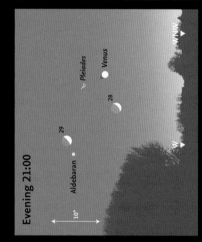

March 31 • Jupiter with Saturn and Mars. Nunki may be found a little more to the south.

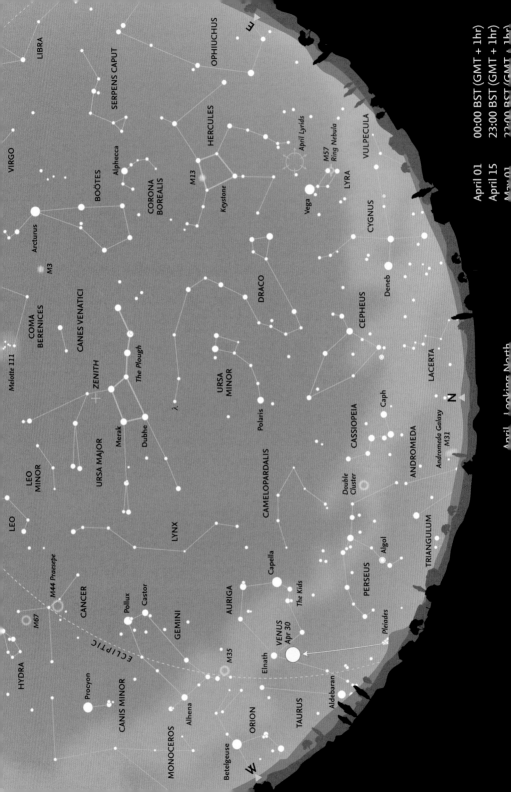

April 01   00:00 BST (GMT + 1hr)
April 15   23:00 BST (GMT + 1hr)
May 01   22:00 BST (GMT + 1hr)

April Looking North

W

E

N

LIBRA

VIRGO

SERPENS CAPUT

OPHIUCHUS

BOÖTES

Arcturus

M3

Alphecca

CORONA
BOREALIS

HERCULES

M13

Keystone

April Lyrids

M57
Ring Nebula

LYRA

Vega

VULPECULA

CYGNUS

Deneb

DRACO

COMA
BERENICES

CANES VENATICI

Melotte 111

The Plough

ZENITH

λ

URSA
MINOR

Polaris

CEPHEUS

Caph

LACERTA

CASSIOPEIA

ANDROMEDA

Andromeda Galaxy
M31

URSA MAJOR

Merak

Dubhe

CAMELOPARDALIS

LEO
MINOR

LEO

LYNX

Double
Cluster

TRIANGULUM

PERSEUS

Algol

Capella

AURIGA

The Kids

M44 Praesepe

CANCER

M67

Pollux

Castor

GEMINI

Alhena

M35

Elnath

VENUS
Apr 30

Pleiades

ECLIPTIC

HYDRA

Procyon

CANIS MINOR

MONOCEROS

Alhena

ORION

Betelgeuse

TAURUS

Aldebaran

April – Looking North

Cygnus and the brighter regions of the Milky Way are now becoming visible, running more-or-less parallel with the horizon in the early part of the night. Rising in the northeast is the small constellation of **Lyra** and the distinctive 'Keystone' of **Hercules** above it. This asterism is very useful for locating the bright globular cluster M13 (see map on page 67), which lies on one side of the quadrilateral. The winding constellation of **Draco** weaves its way from the quadrilateral of stars that marks its 'head', on the border with Hercules, to end at λ Draconis between **Polaris** (α Ursae Minoris) and the 'Pointers', **Dubhe** and **Merak** (α and β Ursae Majoris, respectively). **Ursa Major** is 'upside down' high overhead, near the zenith. The constellation of **Gemini** stands almost vertically in the west. **Auriga** is still clearly seen in the northwest, but, by the end of the month, the southern portion of **Perseus** is starting to dip below the northern horizon. The very faint constellation of **Camelopardalis** lies in the northwest between Polaris and the constellations of Auriga and Perseus.

Meteors

A moderate meteor shower, the **Lyrids**, peaks on April 21–22. Although the hourly rate is not very high (about 18 meteors per hour), the meteors are fast and some leave persistent trains. This year the maximum occurs shortly before New Moon, so conditions are extremely favourable for seeing the fainter meteors. The parent object is the non-periodic comet C/1861 G1 (Thatcher). Another, stronger shower, the **Eta Aquariids**, begins to be active around April 18, and comes to maximum in May.

In April the constellation of Boötes lies halfway between the eastern horizon and the zenith. The brightest star in the photograph is orange-tinted Arcturus. The small constellation of Corona Borealis appears in the upper left-hand corner of the image.

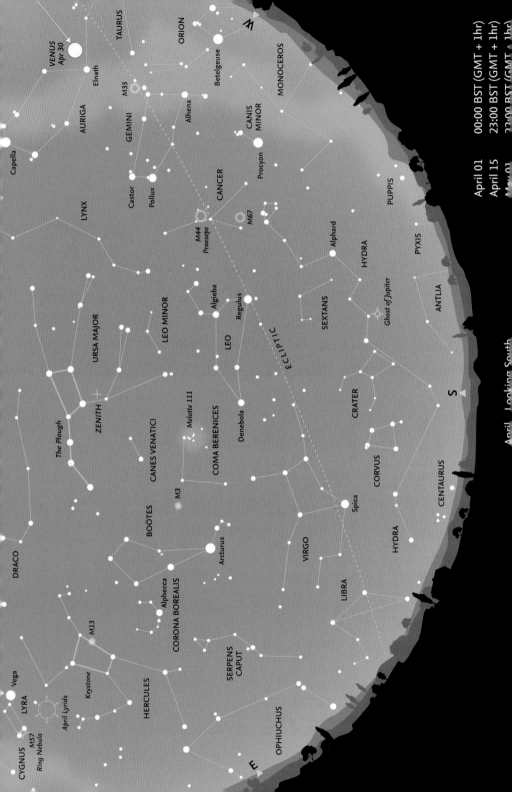

April, Looking South

April 01   00:00 BST (GMT + 1hr)
April 15   23:00 BST (GMT + 1hr)

April – Looking South

Leo is the most prominent constellation in the southern sky in April, and vaguely looks like the creature after which it is named. **Gemini**, with **Castor** and **Pollux**, remains clearly visible in the west, and **Cancer** lies between the two constellations. To the east of Leo, the whole of **Virgo**, with **Spica** (α Virginis) its brightest star, is well clear of the horizon. Below Leo and Virgo, the complete length of **Hydra** (α Hydrae) is visible, running beneath both constellations, with **Alphard** (α Hydrae) halfway between Regulus and the southwestern horizon. Farther east, the two small constellations of **Crater** and the rather brighter **Corvus** lie between Hydra and Virgo.

Boötes and **Arcturus** are prominent in the eastern sky, together with the circlet of **Corona Borealis**, framed by Boötes and the neighbouring constellation of **Hercules**. Between Leo and Boötes lies the constellation of **Coma Berenices**, notable for being the location of the open cluster Melotte 111 (see page 41) and the Coma Cluster of galaxies (Abell 1656). There are about 1,000 galaxies in this cluster, which is located near the North Galactic Pole, where we are looking out of the plane of the Galaxy and are thus able to see deep into space. Only about ten of the brightest galaxies in the Coma Cluster are visible with the largest amateur telescopes.

The distinctive constellation of Leo, with Regulus and 'The Sickle' on the west. Algieba (γ Leonis), north of Regulus, appearing double, is a multiple system of four stars.

The Moon's phases for April

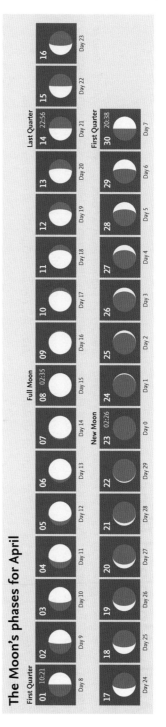

April – Moon and Planets

The Moon

On April 4, the Moon passes 4.0° north of *Regulus*, between it and *Algieba* (γ Leonis). Three days later (April 7) it is at the closest perigee of the year at a distance of 356,900 km. On April 8, the Moon passes north of *Spica* in *Virgo*, and on April 11, a similar distance north of *Antares* in *Scorpius*. On April 15–16 it passes (in succession) *Jupiter*, *Saturn* and *Mars* in the morning sky. It is 3.8° north of *Aldebaran* on April 26.

The planets

Mercury, in *Capricornus* and *Pisces*, is invisible in daylight. *Venus* is in *Taurus*, passing the *Pleiades* on March 2–4, and brightening slightly to mag. -4.7 over the month. *Mars* is in *Capricornus*, brightening from mag. 0.8 to 0.4 during April. *Jupiter* is in *Sagittarius* at mag −2.1 to -2.3. *Saturn* (mag. 0.7–0.6) is just within *Capricornus*. *Uranus* is mag. 5.8 in *Aries*, and *Neptune* (mag. 8.0–7.9) remains in *Aquarius*.

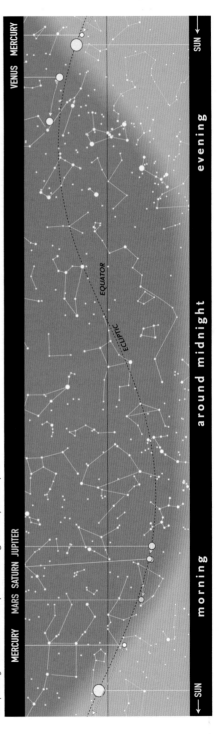

The path of the Sun and the planets along the ecliptic in April.

Calendar for April

01	10:21	First Quarter
02	08:28	Pollux 4.9°N of Moon
04	19:11	Regulus 4.0°S of Moon
07	18:08	Moon at perigee (356,900 km, closest of year)
08	02:35	Full Moon
08	06:48	Spica 7.3°S of Moon
11	11:43	Antares 6.6°S of Moon
13–29		Lyrid meteor shower
14	22:56	Last Quarter
14	23:05	Jupiter 3.0°N of Moon
15	09:18	Saturn 2.5°N of Moon
16	04:33	Mars 3.0°N of Moon
18–May 27		Eta Aquariid meteor shower
20	19:01	Moon at apogee (406,500 km)
21	17:16	Mercury 3.1°N of Moon
22	06:30	Lyrid shower maximum
23	02:26	New Moon
26	04:06	Aldebaran 3.8°S of Moon
26	10:19	Uranus at conjunction with the Sun
26	15:25	Venus 6.1°N of Moon
29	14:49	Pollux 4.7°N of Moon
30	20:38	First Quarter

Evening 21:00 (BST)

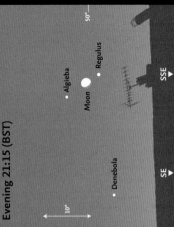

April 2–4 • Venus (mag. –4.5) passes the Pleiades. The brightest member of the cluster (Alcyone) is mag. 2.8.

Morning 5:30 (BST)

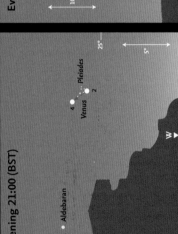

April 15–16 • In the early morning the Moon passes Jupiter, Saturn and Mars, low in the southeast.

Evening 21:15 (BST)

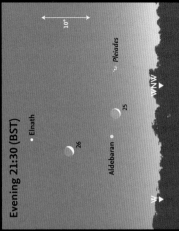

April 4 • The Moon is between Regulus and Algieba, high in the south-southeast.

Evening 21:30 (BST)

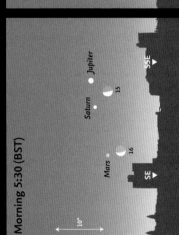

April 25–26 • After sunset the Moon accompanies Aldebaran, the Pleiades and Elnath.

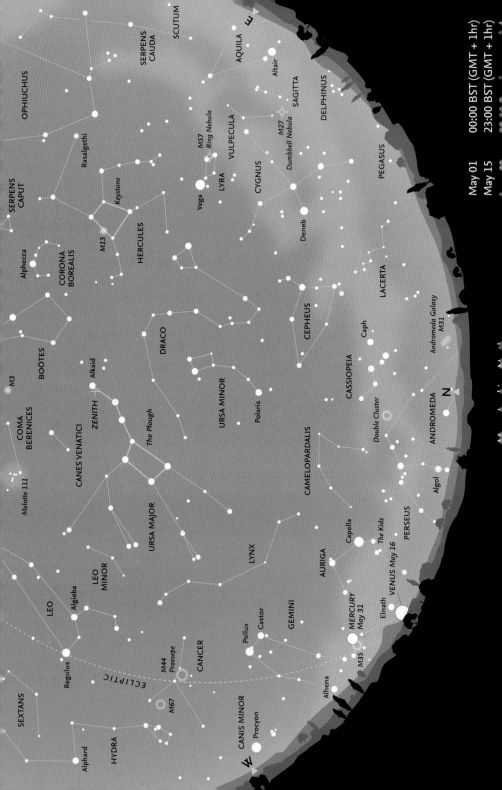

May – Looking North

Cassiopeia is now low over the northern horizon and, to its west, the southern portions of both **Perseus** and **Auriga** are becoming difficult to observe, although the **Double Cluster**, between Perseus and Cassiopeia is still clearly visible. Early in the night, the **Andromeda Galaxy** is too low to be visible with the extinction and other hindrances that occur close to the horizon. The constellations of **Lyra, Cepheus, Ursa Minor** and the whole of **Draco** are well placed in the sky. **Gemini**, with **Castor** and **Pollux**, is sinking towards the western horizon. **Capella** (α Aurigae) and the asterism of **The Kids** are still clear of the horizon.

In the east, two of the stars of the 'Summer Triangle', **Vega** (α Lyrae) and **Deneb** (α Cygni), are clearly visible, and the third star, **Altair** in **Aquila**, is beginning to climb above the horizon. The whole of **Cygnus** is now visible. The sprawling constellation of **Hercules** is high in the east and the brightest globular cluster in the northern hemisphere, M13, is visible to the naked eye on the western side of the asterism known as the **Keystone.**

Three faint constellations may be identified before the lighter nights of summer make them difficult before the lighter northeastern sky is the zig-zag constellation of **Lacerta**, while to the west, above Perseus and Auriga is **Camelopardalis** and, farther west, the line of faint stars forming **Lynx.**

Later in the night (and in the month) the westernmost stars of **Pegasus** begin to come into view, while the stars of **Andromeda** are skimming the northeastern horizon and the Andromeda Galaxy may become visible. High overhead, **Alkaid** (η Ursae Majoris), the last star in the 'tail' of the Great Bear, is close to the zenith, while the main body of the constellation has swung round into the western sky.

Meteors

The **Eta Aquariids** are one of the two meteor showers associated with Comet 1P/Halley (the other being the **Orionids**, in October). The Eta Aquariids are not particularly favourably placed for northern-hemisphere observers, because the radiant is near the celestial equator, near the 'Water Jar' in **Aquarius**, well below the horizon until late in the night (around dawn). However, meteors may still be seen in the eastern sky even when the radiant is below the horizon. There is a radiant map for the Eta Aquariids on page 30.

The maximum in 2020, on May 5–6, occurs just before Full Moon, so conditions are extremely unfavourable and any observations of the shower will be difficult. Maximum hourly rate is about 55 per hour and a large proportion (about 25 per cent) of the meteors leave persistent trains.

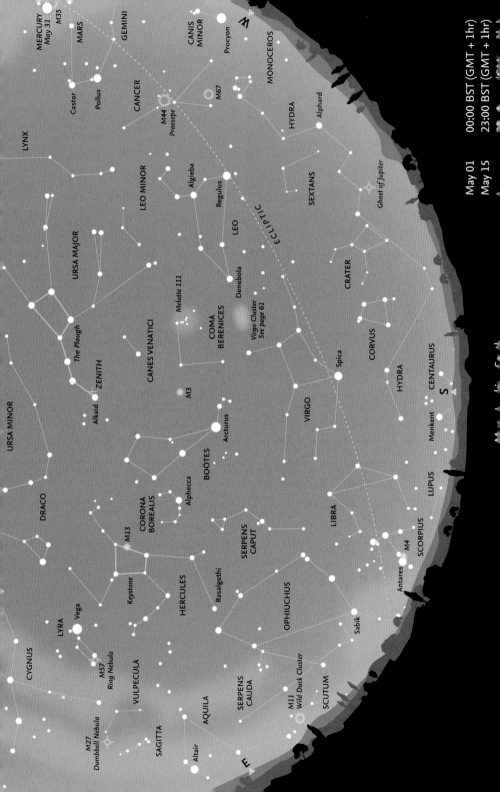

W

MERCURY
May 31
M35
MARS
GEMINI
CANIS
MINOR
Procyon
MONOCEROS
Castor
Pollux
CANCER
M44
Praesepe
HYDRA
Alphard
LYNX
LEO MINOR
Algieba
Regulus
SEXTANS
Ghost of Jupiter
LEO
ECLIPTIC
CRATER
URSA MAJOR
Melotte 111
Denebola
CORVUS
The Plough
CANES VENATICI
COMA
BERENICES
Virgo Cluster
See page 61
HYDRA
ZENITH
Alkaid
M3
Spica
CENTAURUS
URSA MINOR
Arcturus
VIRGO
S
Menkent
DRACO
BOÖTES
Alphecca
LIBRA
LUPUS
CORONA
BOREALIS
M13
SERPENS
CAPUT
SCORPIUS
M4
Keystone
Rasalgethi
Antares
Vega
HERCULES
LYRA
OPHIUCHUS
Sabik
CYGNUS
M57
Ring Nebula
VULPECULA
SERPENS
CAUDA
M11
Wild Duck Cluster
SCUTUM
M27
Dumbbell Nebula
AQUILA
SAGITTA
Altair
E

May – Looking South

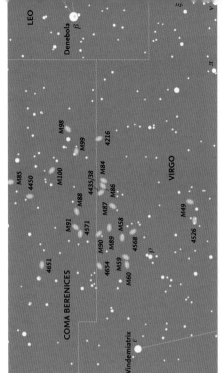

Early in the night, the constellation of *Virgo*, with *Spica* (α Virginis), lies due south, with *Leo* and both *Regulus* and *Denebola* (α and β Leonis, respectively) to its west still well clear of the horizon. Later in the night, the rather faint zodiacal constellation of *Libra* becomes visible and, to its east, the ruddy star *Antares* (α Scorpii) begins to climb up over the horizon.

Virgo contains the nearest large cluster of galaxies, which is the centre of the Local Supercluster, of which the Milky Way galaxy forms part. The Virgo Cluster contains some 2,000 galaxies, the brightest of which are visible in amateur telescopes.

Arcturus in *Boötes* is high in the south, with the distinctive circlet of *Corona Borealis* clearly visible to its east. The brightest star (α Coronae Borealis) is known as *Alphecca*. The large constellation of *Ophiuchus* (which actually crosses the ecliptic, and is thus the 'thirteenth' zodiacal constellation) is climbing into the eastern sky. Before the constellation boundaries were formally adopted by the International Astronomical Union in 1930, the southern region of Ophiuchus was regarded as forming part of the constellation of Scorpius, which had been part of the zodiac since antiquity.

A finder chart for some of the brightest galaxies in the Virgo Cluster (see page 60). All stars brighter than magnitude 8.5 are shown.

The Moon's phases for May

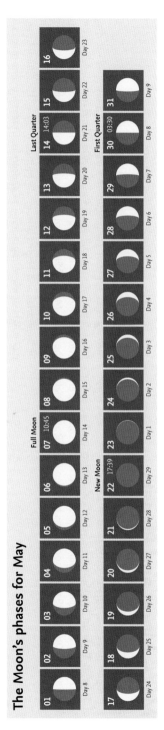

May – Moon and Planets

The Moon

The Moon passes 4.2° north of *Regulus* in *Leo* on May 2, 7.4° north of *Spica* in *Virgo* on May 5 and 6.5° north of *Antares* on May 8. On May 12–15 it passes *Jupiter*, *Saturn* and *Mars*, low in the southeastern morning sky. On May 29 the Moon passes between *Regulus* and *Algieba* (γ Leonis).

The planets

Mercury is at superior conjunction on May 4. On May 20–24 it may be glimpsed at mag. 0.8 to -0.4 as it passes *Venus* (mag. -4.3) in the evening sky. *Venus* is fading (mag. -4.7 to -4.1) as it moves towards the Sun. *Mars* moves from *Capricornus* into *Aquarius*, brightening from mag. 0.4 to -0.0 over the month. *Jupiter* (mag. -2.3 to -2.6) is in *Sagittarius* and begins retrograde motion on May 16. *Saturn* (mag. 0.6–0.4) is slowly retrograding in *Capricornus*. *Uranus* remains mag. 5.8 in *Aries*, and *Neptune* is mag. 7.9 in *Aquarius*.

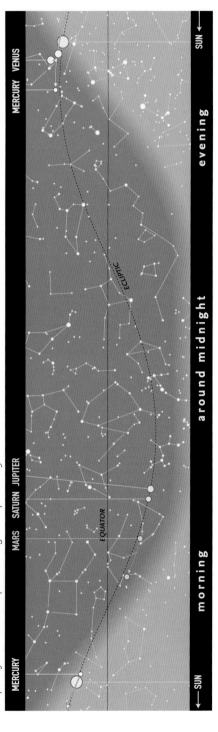

The path of the Sun and the planets along the ecliptic in May.

Calendar for May

02	03:16	Regulus 4.2°S of Moon
04	21:22	Mercury at superior conjunction
05	20:01	Eta Aquariid shower maximum
05	17:24	Spica 7.4°S of Moon
06	03:03	Moon at perigee (359,700 km)
07	10:45	Full Moon
08	22:10	Antares 6.5°S of Moon
12	09:41	Jupiter 2.3°N of Moon
12	18:11	Saturn 2.7°N of Moon
14	14:03	Last Quarter
15	02:02	Mars 2.8°N of Moon
18	07:45	Moon at apogee (405,600 km)
22	08:00 *	Mercury 0.9°S of Venus
22	17:39	New Moon
23	10:11	Aldebaran 3.8°S of Moon
24	02:42	Venus 3.7°N of Moon
24	10:52	Mercury 2.8°N of Moon
26	20:12	Pollux 4.5°N of Moon
29	09:12	Regulus 4.3°S of Moon
30	03:30	First Quarter

* These objects are close together for an extended period around this time.

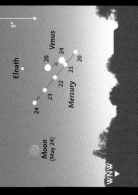

Morning 4:00 (BST)

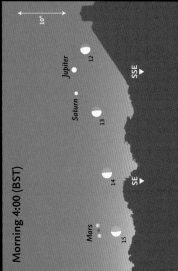

May 12–15 • Low in the southeast, the Moon passes three planets: Jupiter, Saturn and Mars.

Evening 21:45 (BST)

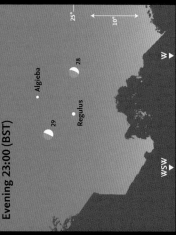

May 20–24 • Mercury passes Venus, and on May 24, a narrow crescent Moon is nearby.

Evening 22:00 (BST)

May 26 • Castor, Pollux, the Moon, Procyon and Alhena almost form a right-angled triangle.

Evening 23:00 (BST)

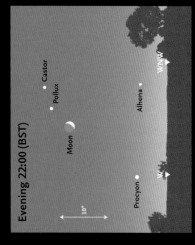

May 28–29 • The Moon passes between Regulus and Algieba in the western sky.

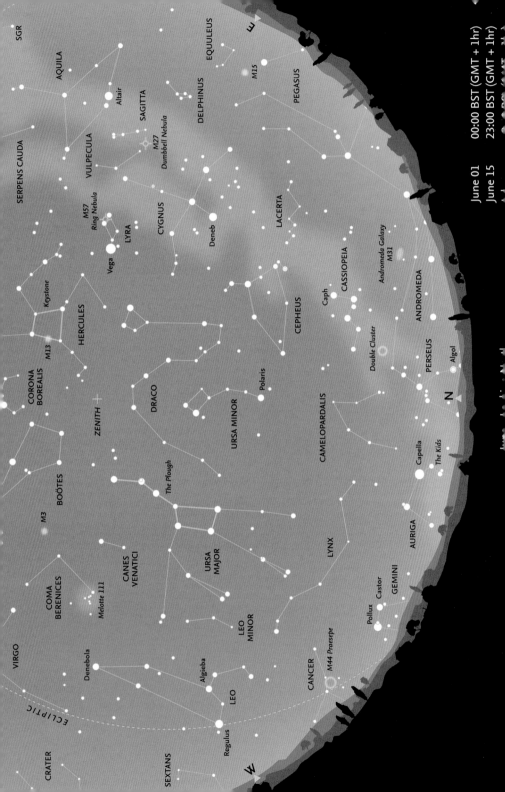

June – Looking North

Around summer solstice (June 20) even in southern England and Ireland a form of twilight persists throughout the night. Farther north, in Scotland, the sky remains so light that most of the fainter stars and constellations are invisible. There, even brighter stars, such as the seven stars making up the well-known asterism known as the *Plough* in *Ursa Major* may be difficult to detect except around local midnight, 00:00 UT (01:00 BST).

But there is one compensation during these light nights: even southern observers may be lucky enough to witness a display of noctilucent clouds (NLC). These are highly distinctive clouds shining with an electric-blue tint, observed in the sky in the direction of the North Pole. They are the highest clouds in the atmosphere, occurring at altitudes of 80–85 km, far above all other clouds. They are only visible during summer nights, for about a month or six weeks on either side of the solstice, when observers are in darkness, but the clouds themselves remain illuminated by sunlight, reaching them from the Sun, itself hidden below the northern horizon.

Noctilucent clouds, photographed on the night of July 5-6, 2016, from Portmahomack, Ross-shire, Scotland, by Denis Buczynski.

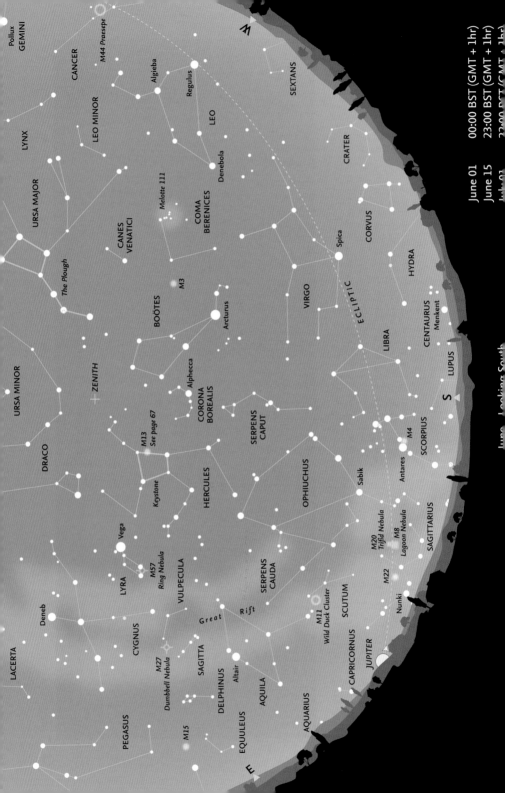

June – Looking South

GEMINI
Pollux

CANCER

M44 Praesepe

Algieba

Regulus

LEO MINOR

LEO

LYNX

SEXTANS

URSA MAJOR

CANES
VENATICI

Melotte 111

CRATER

COMA
BERENICES

Denebola

The Plough

M3

CORVUS

BOÖTES

Spica

HYDRA

Arcturus

VIRGO

ECLIPTIC

URSA MINOR

ZENITH

LIBRA

CENTAURUS

Menkent

DRACO

Alphecca

CORONA
BOREALIS

LUPUS

SERPENS
CAPUT

S

M13
See page 67

Keystone

HERCULES

OPHIUCHUS

SCORPIUS

M4

Antares

Sabik

Vega

LYRA

M20
Trifid Nebula

SAGITTARIUS

M57
Ring Nebula

VULPECULA

M8
Lagoon Nebula

Deneb

SERPENS
CAUDA

M22

Great   Rift

Nunki

CYGNUS

M27
Dumbbell Nebula

M11
Wild Duck Cluster

LACERTA

SCUTUM

SAGITTA

CAPRICORNUS

Altair

DELPHINUS

JUPITER

AQUILA

PEGASUS

EQUULEUS

AQUARIUS

M15

E

June – Looking South

Although the persistent twilight makes observing even the southern sky difficult, the rather undistinguished constellation of **Libra** lies almost due south. The red supergiant star **Antares** – the name means the 'Rival of Mars' – in **Scorpius** is visible slightly to the east of the meridian, but the 'tail' or 'sting' remains below the horizon. Higher in the sky is the large constellation of **Ophiuchus** (the 'Serpent Bearer'), lying between the two halves of the constellation of **Serpens: Serpens Caput** ('Head of the Serpent') to the west and **Serpens Cauda** ('Tail of the Serpent') to the east. (Serpens is the only constellation to be divided into two distinct parts.) The ecliptic runs across Ophiuchus, and the Sun actually spends far more time in the constellation than it does in the 'classical' zodiacal constellation of Scorpius, a small area of which lies between Libra and Ophiuchus.

Higher in the southern sky, the three constellations of **Boötes**, **Corona Borealis** and **Hercules** are now better placed for observation than at any other time of the year. This is an ideal time to observe the fine globular cluster of M13 in Hercules.

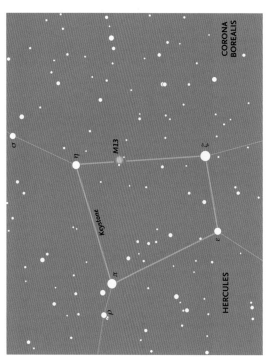

Finder chart for M13, the finest globular cluster in the northern sky. All stars down to magnitude 7.5 are shown.

The Moon's phases for June

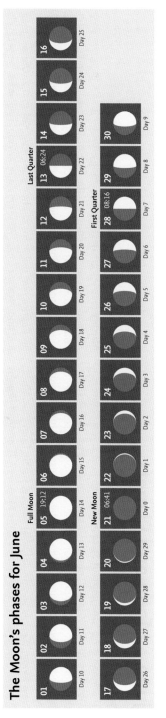

June – Moon and Planets

The Moon

The Moon is 7.6° due north of *Spica* in *Virgo* on June 2. Just before Full Moon on June 5 it is 6.5° north of *Antares* in *Scorpius*. At Full Moon there is a faint penumbral lunar eclipse. The Moon passes south of *Jupiter* and *Saturn* on June 8–9, low in the southeast. On June 12, just before Last Quarter, the Moon is 2.8° south of *Mars*. On June 21 at New Moon, there is an annular solar eclipse, visible from East Africa, Arabia, India and China. On June 25, the Moon is 4.7° north of *Regulus* and 7.8° north of *Spica* again on June 29.

The planets

Mercury reaches greatest eastern elongation on June 4 (diagram page 22), and is close to Venus in late May in the evening after sunset. *Venus* is invisible at inferior conjunction on June 3. *Mars* (mag. -0.0 to -0.5) moves from *Aquarius* into *Pisces. Jupiter* (mag. -2.6 to -2.7) continues retrograde motion in *Sagittarius. Saturn*, also retrograding (in *Capricornus*) brightens slightly from mag. 0.4 to 0.2. *Uranus* (mag. 5.8) and *Neptune* (mag. 7.9) remain in *Aries* and *Aquarius*, respectively. Neptune begins retrograde motion on June 24. Minor planet *(7) Iris* reaches opposition (mag. 8.8) on June 28.

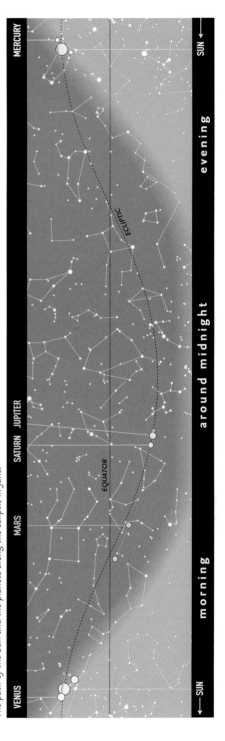

The path of the Sun and the planets along the ecliptic in June.

Calendar for June

02	01:57	Spica 7.6°S of Moon
03	03:36	Moon at perigee (364,400 km)
03	17:42	Venus at inferior conjunction
04	12:59	Mercury at greatest elongation (23.6°E, mag. 0.4)
05	08:20	Antares 6.5°S of Moon
05	19:12	Full Moon
05	19:25	Penumbral lunar eclipse
08	17:21	Jupiter 2.2°N of Moon
09	02:12	Saturn 2.7°N of Moon
12	23:55	Mars 2.8°N of Moon
13	06:24	Last Quarter
19	17:28	Aldebaran 3.9°S of Moon
19	08:54	Venus 0.7°S of Moon
20	21:44	Summer solstice
21	06:40	Annular solar eclipse
21	06:41	New Moon
22	07:18	Mercury 3.9°S of Moon
23	02:22	Pollux 4.5°N of Moon
25	14:36	Regulus 4.7°S of Moon
28	01:48	Minor planet (7) Iris at opposition (mag. 8.8)
28	08:16	First Quarter
29	08:16	Spica 7.8°S of Moon
30	02:09	Moon at perigee (369,000 km)

Evening 23:00 (BST)

June 4–5 • *The Full Moon passes close by Acrab (β Sco). One day later it has passed Antares and is below Sabik (η Oph).*

After midnight 1:00 (BST)

June 2 • *The Moon with Spica in the southwest, shortly after midnight.*

After midnight 1:00 (BST)

June 9 • *The Moon with Jupiter and Saturn, low in the southeast. Nunki may be found farther to the south.*

Early morning 3:00 (BST)

June 13 • *The Last Quarter Moon is just below Mars.*

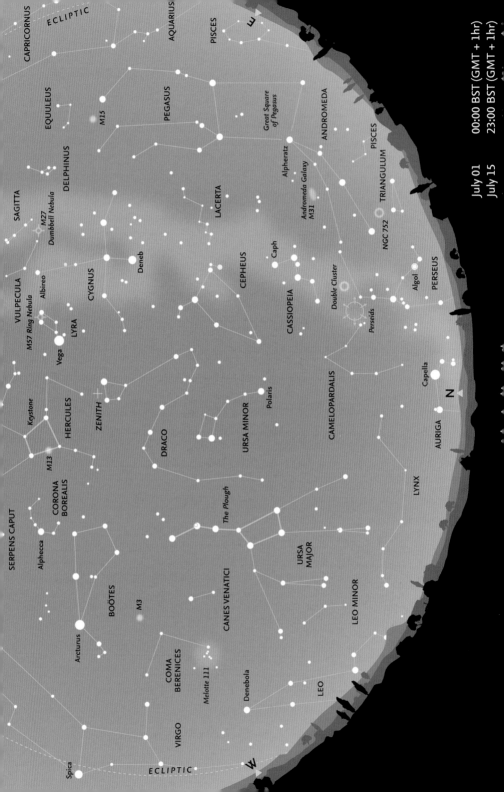

July – Looking North

As in June, light nights and the chance of observing noctilucent clouds persist throughout July, but later in the month (and particularly after midnight) some of the major constellations begin to be more easily seen. **Capella**, the brightest star in **Auriga** (most of which is too low to be visible), is skimming the northern horizon. **Cassiopeia** is clearly visible in the northeast and **Perseus**, to its south, is beginning to climb clear of the horizon. The band of the Milky Way, from Perseus through Cassiopeia towards **Cygnus**, stretches up into the northeastern sky. If the sky is dark and clear, you may be able to make out the small, faint constellation of **Lacerta**, lying across the Milky Way between Cassiopeia and Cygnus. In the east, the stars of **Pegasus** are now well clear of the horizon, with the main line of stars forming **Andromeda** roughly parallel to the horizon in the northeast. **Alpheratz** (α Andromedae) is actually the star at the northeastern corner of the **Great Square of Pegasus**. **Cepheus** and **Ursa Major** are on opposite sides of **Polaris** and **Ursa Minor**, in the east and west, respectively. The head of **Draco** is very close to the zenith so the whole of this winding constellation is readily seen.

Meteors

July brings increasing meteor activity, mainly because there are several minor radiants active in the constellations of **Capricornus** and **Aquarius**. Because of their location, however, observing conditions are not particularly favourable for northern-hemisphere observers, although the first shower, the **Alpha Capricornids**, active from July 2 to August 14 (peaking July 29, with a tail to August 14), does often produce very bright fireballs. The maximum rate, however, is only about 5 per hour. The parent body is Comet 169P/NEAT. The most prominent shower is probably that of the **Delta Aquariids**, which are active from around July 13 to August 24, with a peak on July 30–Aug 1, although even then the hourly rate is unlikely to reach 20 meteors per hour. In this case, the

Cygnus, sometimes known as the 'Northern Cross', depicts a swan flying down the Milky Way towards Sagittarius. The brightest star, Deneb (α Cygni), represents the tail, and Albireo (β Cygni) marks the position of the head, and may be found at bottom-right of this image.

parent body is possibly Comet 96P/Machholz. This year both shower maxima occur when the Moon is a waxing gibbous, so observing conditions are moderately unfavourable. A chart showing the **Delta Aquariid** radiant is shown on page 30. The **Perseids** begin on July 16 and peak on August 11–12.

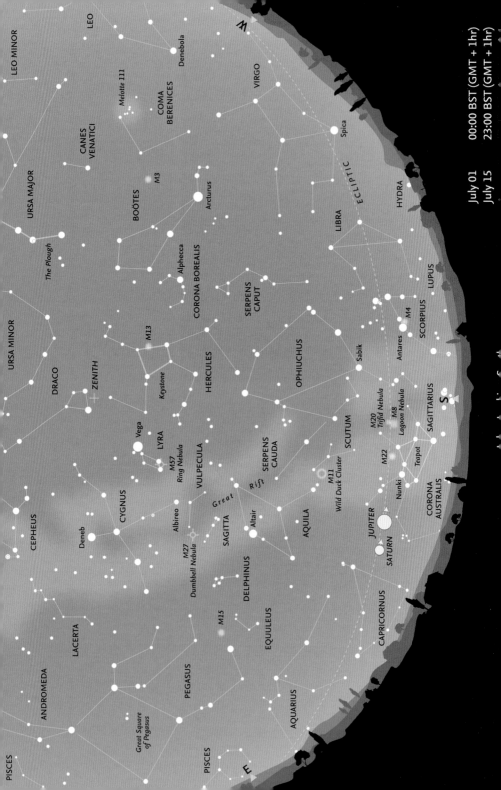

LEO MINOR

LEO

LEO

VIRGO

Denebola

Melotte 111

COMA
BERENICES

CANES
VENATICI

Spica

Arcturus

M3

BOÖTES

URSA MAJOR

ECLIPTIC

LIBRA

HYDRA

The Plough

Alphecca

CORONA BOREALIS

SERPENS
CAPUT

LUPUS

M4

URSA MINOR

DRACO

ZENITH

M13

HERCULES

OPHIUCHUS

Sabik

Antares

SCORPIUS

Vega

LYRA

Keystone

SCUTUM

M20
Trifid Nebula

M8
Lagoon Nebula

SAGITTARIUS

CEPHEUS

Deneb

CYGNUS

M57
Ring Nebula

VULPECULA

SERPENS
CAUDA

Great Rift

M11
Wild Duck Cluster

M22

Nunki

Teapot

CORONA
AUSTRALIS

Albireo

SAGITTA

Altair

AQUILA

JUPITER

SATURN

LACERTA

M27
Dumbbell Nebula

DELPHINUS

EQUULEUS

CAPRICORNUS

ANDROMEDA

PEGASUS

M15

Great Square
of Pegasus

PISCES

AQUARIUS

PISCES

July – Looking South

Although part of the constellation remains hidden, this is perhaps the best time of year to see *Scorpius*, with deep red *Antares* (α Scorpii), glowing just above the southern horizon. At around midnight (UT), 01:00 BST, part of *Sagittarius*, with the distinctive asterism of the 'Teapot', and the dense star clouds of the centre of the Milky Way, are just visible in the south. The *Great Rift* – actually dust clouds that hide the more distant stars – runs down the Milky Way from Cygnus towards Sagittarius. Towards its northen end is the small constellation of *Sagitta* and the planetary nebula *M27* (the Dumbbell Nebula). The sprawling constellation of *Ophiuchus* lies close to the meridian for a large part of the month, separating the two halves of the constellation of *Serpens*. The western half is called *Serpens Caput* (Head of the Serpent) and the eastern part *Serpens Cauda* (Tail of the Serpent). In the east, the bright *Summer Triangle*, consisting of *Vega* in *Lyra*, *Deneb* in *Cygnus* and *Altair* in *Aquila*, begins to dominate the southern sky, as it will throughout August and into September. The small constellation of Lyra, with Vega and a distinctive quadrilateral of stars to its east and south, lies not far south of the zenith.

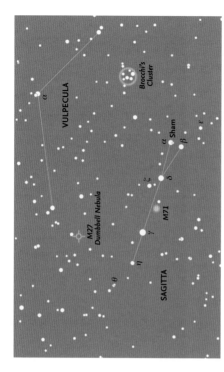

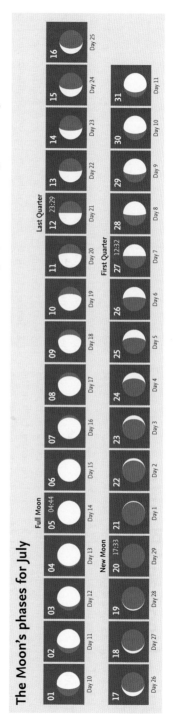

A finder chart for M27, the Dumbbell Nebula, a relatively bright (magnitude 8) planetary nebula – a shell of material ejected in the late stages of a star's lifetime – in the constellation of Vulpecula. All stars brighter than magnitude 7.5 are shown.

The Moon's phases for July

				Full Moon				
01	02	03	04	05 04:44	06	07	08	09
Day 9	Day 10	Day 11	Day 12	Day 13	Day 14	Day 15	Day 16	Day 17

New Moon

10	11	12 23:29	13	14	15	16
Day 18	Day 19	Day 20	Day 21	Day 22	Day 23	Day 24	Day 25

Last Quarter

First Quarter

July – Moon and Planets

The Earth

The Earth reaches *aphelion* (the farthest point from the Sun) on July 4 at 11:35 UT, when its distance is 1.0167 AU (= 152,095,223 km).

The Moon

On July 2, the Moon passes 6.5° north of **Antares**. On July 5 there is yet another faint penumbral lunar eclipse. That day (at Full Moon) it passes 1.9° south of **Jupiter** and on July 6, 2.5° south of **Saturn**, both in the south-southeastern sky. On July 11, **Mars** (in **Pisces**) is just 2.0° north of the Moon. On July 17, the Moon is 3.9° north of **Aldebaran** (α Tauri) and later that day, 3.1° north of **Venus**. It is 4.3° north of **Regulus** on July 22. On July 26, the Moon is 7.4° north of **Spica** in **Virgo** in the evening sky and on July 29, 6.4° north of **Antares**.

The planets

Mercury is at inferior conjunction on July 1 but comes to greatest western elongation on July 22, when Venus is higher above it in the morning sky (diagram page 22). **Venus** is in **Taurus**, moving towards the Sun, and very bright (mag. -4.7 to -4.6). It is just 1.0° north of **Aldebaran** on July 12. **Mars** is in **Pisces** and continues to brighten (mag. -0.5 to -1.1). On July 14, **Jupiter** is at opposition in **Sagittarius**, at mag. -2.8. **Saturn**, still slowly retrograding, moves into **Sagittarius** and also comes to opposition (on July 20) at mag. 0.1. **Uranus** (mag. 5.8) is in **Aries** and **Neptune** (mag. 7.9) in **Aquarius**.

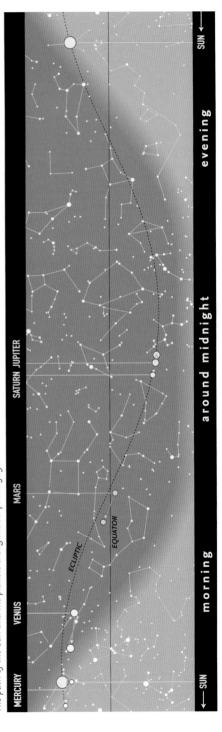

The path of the Sun and the planets along the ecliptic in July.

Date	Time	Event
1	02:45	Mercury at inferior conjunction
2-Aug. 14		Alpha Capricornid meteor shower
2	16:46	Antares 6.5°S of Moon
4	11:35	Earth at aphelion (152,095,223 km = 1.0167 AU)
5	04:30	Penumbral lunar eclipse
5	04:44	Full Moon
5	21:39	Jupiter 1.9°N of Moon
6	08:38	Saturn 2.5°N of Moon
1	19:38	Mars 2°N of Moon
2	07:00 *	Venus 1°N of Aldebaran
2	19:27	Moon at apogee (404,200 km)
2	23:29	Last Quarter
3-Aug. 24		Delta Aquariid meteor shower
4	07:03	Jupiter at opposition (mag. −2.8)
6-Aug. 23		Perseid meteor shower
7	01:53	Aldebaran 3.9°S of Moon
9	07:26	Venus 3.1°S of Moon
9	03:54	Mercury 3.9°S of Moon
0	10:16	Pollux 4.5°N of Moon
0	17:33	New Moon
0	21:33	Saturn at opposition (mag. 0.1)
2	14:59	Mercury at greatest elongation (20.1°W, mag. 0.3)
2	21:17	Regulus 4.3°S of Moon
5	04:54	Moon at perigee (368,400 km)
6	13:42	Spica 7.4°S of Moon
7	12:32	First Quarter
2	23:10	Antares 6.4°S of Moon
9	23:24	Alpha Capricornid shower maximum
1	23:23	Delta Aquariid shower maximum

These objects are close together for an extended period of this time.

Early morning 4:00 (BST)

10°

—30°

Mars

11

12

Diphda •

SE

SSE

July 11–12 • The Moon with Mars. Diphda (β Cet) lies 15 degrees closer to the horizon.

Early morning 3:00 (BST)

10°

7

Saturn

6

Jupiter

5

Nunki •

S

SSW

July 5–7 • The Full Moon passes Nunki, Jupiter and Saturn, in the early morning low in the south-southwest.

Early morning 4:00 (BST)

10°

Pleiades

Moon

Venus •

Aldebaran

ENE

E

July 12 • In the early morning Venus is close to Aldebaran

Early morning 4:00 (BST)

10°

Pleiades

Venus •

Aldebaran

ENE

E

July 17 • Five days later the Moon is close to Venus and Aldebaran

Early morning 4:30 (BST)

10°

Elnath •

Moon

• Mercury

July 19 • The narrow crescent Moon with Mercury shortly before sunrise

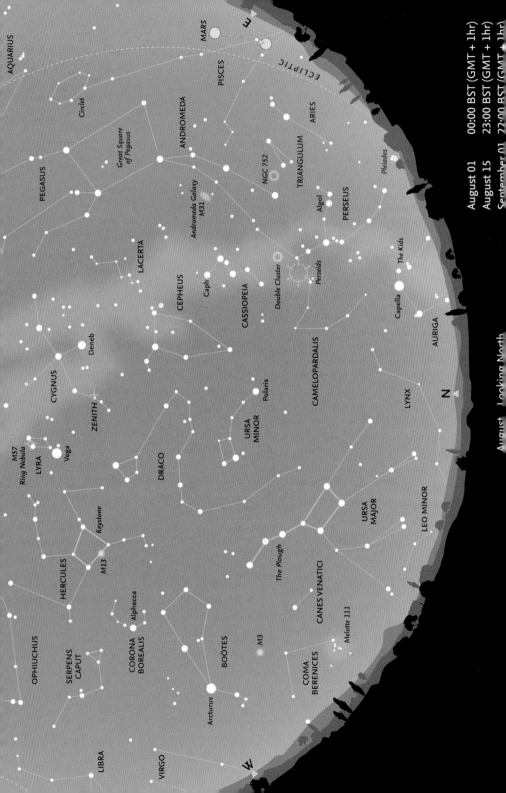

AQUARIUS

MARS
E
PISCES
ECLIPTIC
ANDROMEDA
ARIES
Circlet
PEGASUS
Great Square
of Pegasus
NGC 752
TRIANGULUM
Pleiades
Andromeda Galaxy
M31
Algol
PERSEUS
LACERTA
Double Cluster
Perseids
The Kids
CEPHEUS
Caph
CASSIOPEIA
Capella
Deneb
AURIGA
CYGNUS
CAMELOPARDALIS
N
ZENITH
Polaris
LYNX
M57
Ring Nebula
LYRA
Vega
URSA
MINOR
DRACO
Keystone
URSA
MAJOR
HERCULES
M13
LEO MINOR
The Plough
OPHIUCHUS
CORONA
BOREALIS
Alphecca
BOÖTES
CANES VENATICI
SERPENS
CAPUT
M3
Melotte 111
COMA
BERENICES
LIBRA
Arcturus
W
VIRGO

August 01   00:00 BST (GMT + 1hr)
August 15   23:00 BST (GMT + 1hr)
September 01 22:00 BST (GMT + 1hr)

August · Looking North

August – Looking North

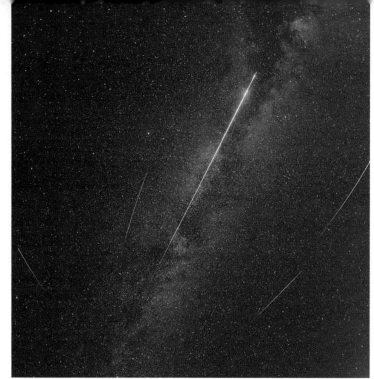

A brilliant Perseid fireball, streaking alongside the Great Rift in the Milky Way, photographed in 2012 by Jens Hackmann from near Weikersheim in Germany. Four additional, fainter Perseids are also visible in the image.

Ursa Major is now the 'right way up' in the northwest, although some of the fainter stars in the south of the constellation are difficult to see. Beyond it, *Boötes* stands almost vertically in the west, but pale orange *Arcturus* is sinking towards the horizon. Higher in the sky, both *Corona Borealis* and *Hercules* are clearly visible.

In the northeast, *Capella* is clearly seen, but most of *Auriga* still remains below the horizon. Higher in the sky, *Perseus* is gradually coming into full view and, later in the night and later in the month, the beautiful *Pleiades* cluster rises above the northeastern horizon. Between Perseus and *Polaris* lies the faint and unremarkable constellation of *Camelopardalis*.

Higher still, both *Cassiopeia* and *Cepheus* are well placed for observation, despite the fact that Cassiopeia is completely immersed in the band of the Milky Way, as is the 'base' of Cepheus. *Pegasus* and *Andromeda* are now well above the eastern horizon and, below them, the constellation of *Pisces* is climbing into view. Two of the stars in the *Summer Triangle*, *Deneb* and *Vega*, are close to the zenith high overhead.

Meteors

August is the month when one of the best meteor showers of the year occurs: the *Perseids*. This is a long shower, generally beginning about July 16 and continuing until around August 23, with a maximum in 2020 on August 11–12, when the rate may reach as high as 100 meteors per hour (and on rare occasions, even higher). In 2020, maximum is around Last Quarter, so conditions are moderately favourable. The Perseids are debris from Comet 109P/Swift-Tuttle (the Great Comet of 1862). Perseid meteors are fast and many of the brighter ones leave persistent trains. Some bright fireballs also occur during the shower.

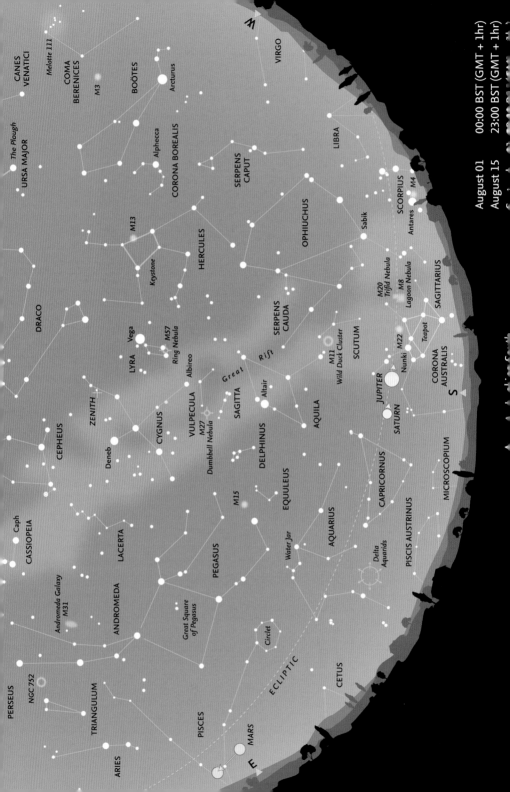

August – Looking South

The whole stretch of the summer Milky Way stretches across the sky in the south, from **Cygnus**, high in the sky near the zenith, past **Aquila**, with bright **Altair** (α Aquilae), to part of the constellation of **Sagittarius** close to the horizon, where the pattern of stars known as the 'Teapot' may be just visible. This area contains many nebulae and both open and globular clusters. Between **Albireo** (β Cygni) and Altair lie the two small constellations of **Vulpecula** and **Sagitta**, with the latter easier to distinguish (because of its shape) from the clouds of the Milky Way. Between Sagitta and **Pegasus** to the east lie the highly distinctive five stars that form the tiny constellation of **Delphinus** (again, one of the few constellations that actually bear some resemblance to the creatures after which they are named). Below Aquila, mainly in the star clouds of the Milky Way, lies **Scutum**, most famous for the bright open cluster, **M11** or the 'Wild Duck Cluster', readily visible in binoculars. To the southeast of Aquila lie the two zodiacal constellations of **Capricornus** and **Aquarius**.

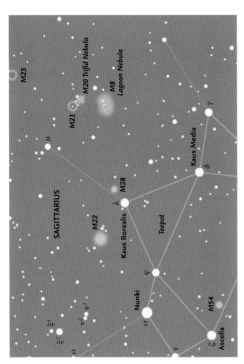

A finder chart for the gaseous nebulae M8 (the Lagoon Nebula) and M20 (the Trifid Nebula) and the globular cluster M22, all in Sagittarius. Clusters M21 & M23 (open) and M28 (globular) are faint. The chart shows all stars brighter than magnitude 7.5.

The Moon's phases for August

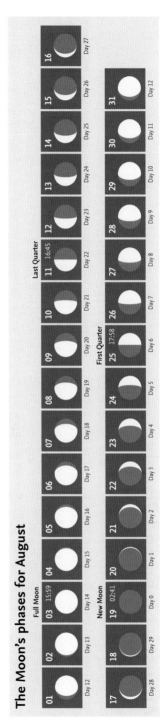

August – Moon and Planets

The Moon

On August 1–2 in **Sagittarius**, the Moon passes 1.5° south of **Jupiter** and then 2.3° south of **Saturn**. On August 9 it is 0.8° south of **Mars** in **Pisces**. Four days later, on August 13, it is 4.0° north of **Aldebaran** in **Taurus**. On August 15 in the morning sky it passes 4.0° north of **Venus**. On August 19 it is 4.3° north of **Regulus** in **Leo**. One week later, on August 26, it passes 6.2° north of **Antares**. On August 29, for the second time in the month, it passes 1.4° south of **Jupiter** and 2.2° south of **Saturn**.

The planets

Mercury is close to the Sun and passes superior conjunction on August 17. **Venus** is an early-morning object, passing from **Taurus** into **Gemini** as it approaches the Sun. It is at greatest elongation (45.8°W) on August 13 (diagram page 22), when it is mag. −4.4 and remains clearly visible into September. **Mars** is in **Pisces**, brightening from mag. −1.1 to −1.8 over the month. **Jupiter** is still slowly retrograding in **Sagittarius** at mag. −2.7 to −2.6. **Saturn**, also retrograding in **Sagittarius**, fades slightly from mag. 0.1 to 0.3. **Uranus** (mag. 5.8) and **Neptune** (mag. 7.9) remain in **Aries** and **Aquarius** respectively. **Uranus** begins retrograde motion on August 16. Dwarf planet (**1**) **Ceres** is at opposition in Aquarius on August 28 at mag. 7.7.

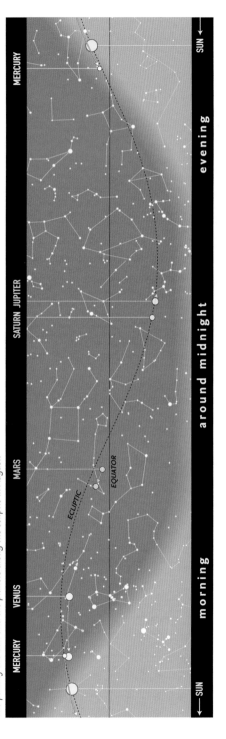

The path of the Sun and the planets along the ecliptic in August.

Calendar for August

1	23:33	Jupiter 1.5°N of Moon
2	13:10	Saturn 2.3°N of Moon
2	06:00 ✱	Mercury 6.6°S of Pollux
3	15:59	Full Moon
9	08:00	Mars 0.8°N of Moon
9	13:51	Moon at apogee (404,700 km)
11	16:45	Last Quarter
12	13:11	Perseid shower maximum
13	00:59	Venus at greatest elongation (45.8°W, mag. –4.4)
13	10:39	Aldebaran 4.0°S of Moon
15	13:00	Venus 4.0°S of Moon
16	19:39	Pollux 4.5°N of Moon
17	14:47	Mercury at superior conjunction
19	02:41	New Moon
19	03:46	Mercury 2.8°S of Moon
19	06:02	Regulus 4.3°S of Moon
21	10:59	Moon at perigee (363,500 km)
22	20:13	Spica 7.2°S of Moon
25	17:58	First Quarter
26	04:36	Antares 6.2°S of Moon
28–Sep. 05		Alpha Aurigid meteor shower
28		Dwarf planet (1) Ceres at opposition (mag. 7.7)
29	01:36	Jupiter 1.4°N of Moon
29	16:32	Saturn 2.2°N of Moon
30	16:18	Venus 2.9°S of Moon
31	20:27	Alpha Aurigid shower maximum

These objects are close together for an extended

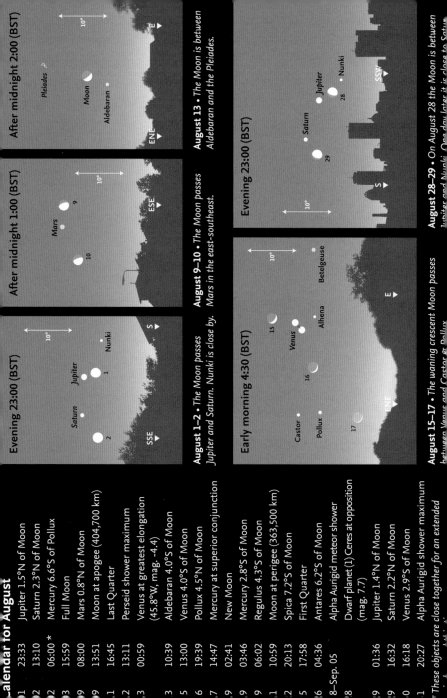

Evening 23:00 (BST)

After midnight 1:00 (BST)

After midnight 2:00 (BST)

August 1–2 • *The Moon passes Jupiter and Saturn. Nunki is close by.*

August 9–10 • *Mars in the east-southeast.*

August 13 • *The Moon is between Aldebaran and the Pleiades.*

Early morning 4:30 (BST)

Evening 23:00 (BST)

August 15–17 • *The waning crescent Moon passes between Venus and Castor & Pollux.*

August 28–29 • *On August 28 the Moon is between Jupiter and Nunki. One day later it is close to Saturn.*

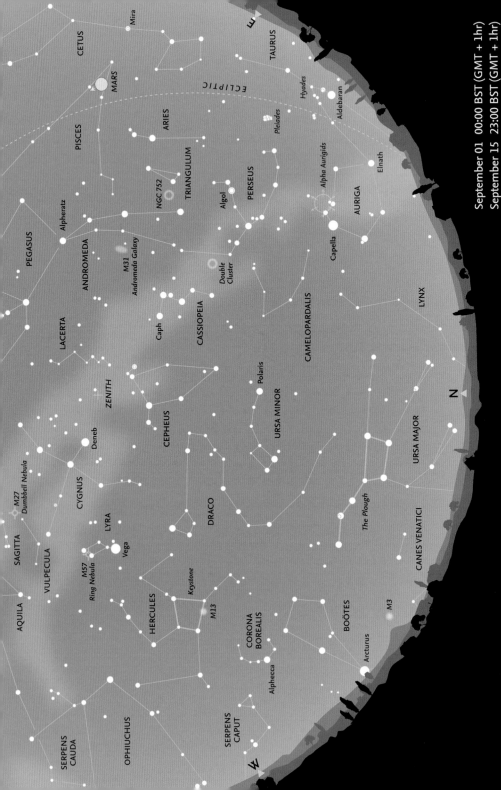

September | Looking North

September – Looking North

The twin open clusters, known as the Double Cluster, in Perseus (more formally called h and χ Persei) are close to the main portion of the Milky Way.

Ursa Major is now low in the north and to the northwest *Arcturus* and much of *Boötes* sink below the horizon later in the night and later in the month. In the northeast *Auriga* is beginning to climb higher in the sky. Later in the month, *Taurus*, with orange *Aldebaran* (α Tauri), and even *Gemini*, with *Castor* and *Pollux*, become visible in the east and northeast. Due east, *Andromeda* is now clearly visible, with the small constellations of *Triangulum* and *Aries* (the latter a zodiacal constellation) directly below it. Practically the whole of the Milky Way is visible, arching across the sky, both in the north and in the south. It is not particularly clear in Auriga, or even *Perseus*, but in *Cassiopeia* and on towards *Cygnus* the clouds of stars become easier to see. The *Double Cluster* in Perseus is well placed for observation. Cepheus is 'upside-down' near the zenith, and the head of *Draco* and *Hercules* (with the globular cluster *M13*), beyond it are well placed for observation.

Meteors

After the major Perseid shower in August, there is very little shower activity in September. One minor, but very extended, shower, known as the *Alpha Aurigids*, actually has two peaks of activity. The first may be on August 28, but the primary peak occurs on August 31. At either of the maxima, however, the hourly rate hardly reaches 10 meteors per hour, although the meteors are bright and relatively easy to photograph. Activity from this shower also extends into October. The *Southern Taurid* shower begins this month and, although rates are low, often produces very bright fireballs. As a slight compensation for the lack of activity, however, in September the number of sporadic meteors reaches its highest rate than at any other time during the year.

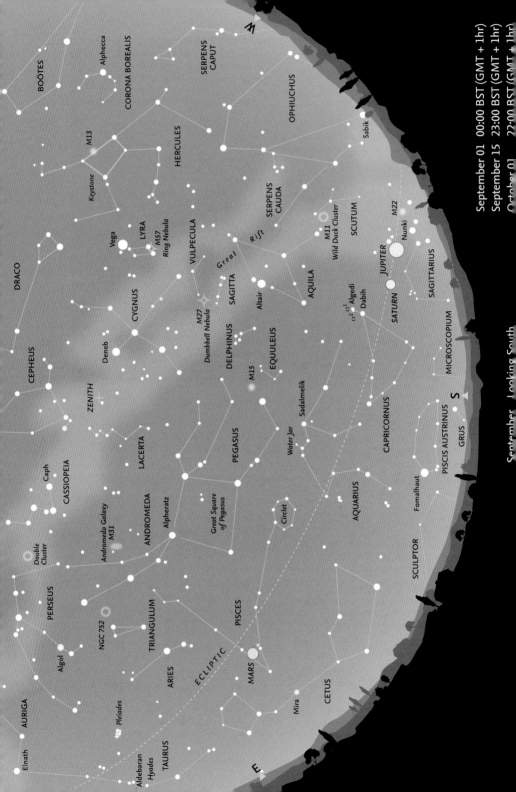

September 01 00:00 BST (GMT + 1hr)
September 15 23:00 BST (GMT + 1hr)
October 01 22:00 BST (GMT + 1hr)

September Looking South

BOOTES
Alphecca
CORONA BOREALIS
SERPENS CAPUT
OPHIUCHUS
Sabik
M13
Keystone
HERCULES
DRACO
SERPENS CAUDA
Great Rift
SAGITTA
VULPECULA
LYRA
Vega
M57
Ring Nebula
M27
Dumbbell Nebula
CYGNUS
Deneb
CEPHEUS
ZENITH
LACERTA
CASSIOPEIA
Caph
Andromeda Galaxy
M31
ANDROMEDA
Alpheratz
Great Square
of Pegasus
PEGASUS
M15
EQUULEUS
DELPHINUS
Altair
AQUILA
α² α¹
Algedi
Dabih
M11
Wild Duck Cluster
SCUTUM
JUPITER
SATURN
M22
Nunki
SAGITTARIUS
MICROSCOPIUM
CAPRICORNUS
Sadalmelik
Water Jar
AQUARIUS
PISCIS AUSTRINUS
GRUS
S
Fomalhaut
SCULPTOR
Circlet
PISCES
ECLIPTIC
MARS
CETUS
Mira
ARIES
TRIANGULUM
NGC 752
Algol
PERSEUS
Double
Cluster
Pleiades
AURIGA
Elnath
Aldebaran
Hyades
TAURUS
E

September – Looking South

The **Summer Triangle** is now high in the southwest, with the Great Square of **Pegasus** high in the southeast. Below Pegasus are the two zodiacal constellations of **Capricornus** and **Aquarius**. In what is otherwise an unremarkable constellation, **Algedi** (α Capricorni) is actually a visual binary, with the two stars (α¹ Cap and α² Cap) readily seen with the naked eye. **Dabih** (β Capricorni), just to the south, is also a double star, and the components are relatively easy to separate with binoculars. In Aquarius, just to the east of **Sadalmelik** (α Aquarii) there is a small asterism consisting of four stars, resembling a tiny letter 'Y', known as the 'Water Jar'. Below Aquarius is a sparsely populated area of the sky with just one bright star in the constellation of **Piscis Austrinus.** In classical illustrations, water is shown flowing from the 'Water Jar' towards bright **Fomalhaut** (α Piscis Austrini).

Another zodiacal constellation, **Pisces,** is now clearly visible to the east of Aquarius. Although faint, there is a distinctive asterism of stars, known as the 'Circlet', south of the Great Square and another line of faint stars to the east of Pegasus. Still farther down towards the horizon

Aquila, with Altair (α Aquilae), its brightest star, like Cygnus, represents a bird – in this case, an eagle – flying down the length of the Milky Way.

is the constellation of **Cetus,** with the famous variable star **Mira** (o Ceti) at its centre. When Mira is at maximum brightness (around mag. 3.5) it is clearly visible to the naked eye, but it disappears as it fades towards minimum (about mag. 9.5 or less). There is a finder chart for Mira on page 97.

The Moon's phases for September

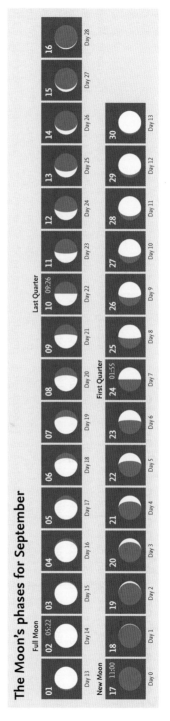

September – Moon and Planets

The Moon

On September 6, the Moon passes just north (0.1°) of **Mars**. An occultation of the planet is visible from a large region of South America. On September 9, the Moon passes 4.2° north of **Aldebaran** in **Taurus**. The Moon passes south of **Castor** and **Pollux** on September 13, and the next day (September 14) it passes 4.5° north of **Venus**. On September 19, the Moon is 7.0° north of **Spica** in **Virgo**, and on September 22, 5.9° north of **Antares**. On September 25 it passes **Jupiter** and, 14 hours later, **Saturn**, at mag. -2.4 and 0.4, respectively.

The planets

Mercury is too close to the Sun to be readily visible this month. **Venus** is a prominent object in the morning sky at mag. -4.3 to -4.1. **Mars**, in **Pisces**, brightens from mag. -1.8 to -2.5 over the month. It begins retrograde motion on September 11. **Jupiter** (in **Sagittarius**) resumes direct motion on September 13. It fades slightly from mag. -2.6 to -2.4 over the month. **Saturn** resumes direct motion at the very end of the month (on September 30) over which it fades slightly (from mag. 0.3 to 0.5). **Uranus** is still in **Aries** at mag. 5.8. **Neptune** comes to opposition in **Aquarius** on September 11, when it is mag. 7.8.

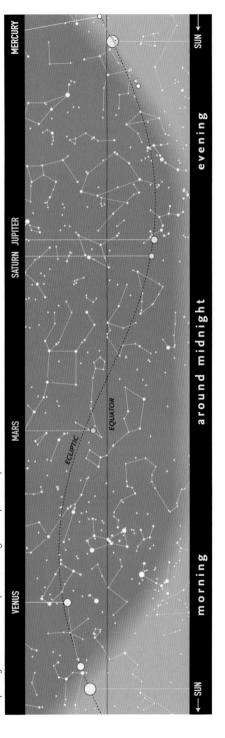

The path of the Sun and the planets along the ecliptic in September.

Calendar for September

01	17:00 *	Venus 1°S of Pollux
02	05:22	Full Moon
06	04:46	Mars 0.1°S of Moon (occultation over S. America)
06	06:31	Moon at apogee (405,600 km)
09	18:45	Aldebaran 4.2°S of Moon
10–Nov. 20		Southern Taurid meteor shower
10	09:26	Last Quarter
11	19:15	Neptune at opposition (mag. 7.8)
13	05:20	Pollux 4.3°N of Moon
14	04:43	Venus 4.5°S of Moon
15	16:17	Regulus 4.3°S of Moon
17	11:00	New Moon
18	13:44	Moon at perigee (359,100 km)
18	21:53	Mercury 6.4°S of Moon
19	05:02	Spica 7.0°S of Moon
22	10:58	Antares 5.9°S of Moon
22	13:31	Northern autumnal equinox
24	01:55	First Quarter
25	06:48	Jupiter 1.6°N of Moon
25	20:38	Saturn 2.3°N of Moon

These objects are close together for an extended period around this time.

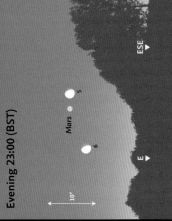

Evening 23:00 (BST)

Mars 5

6

E ▸

ESE ▸

10°

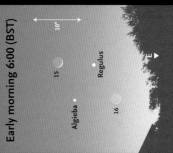

After midnight 0:30 (BST)

⁑ Pleiades

9

Aldebaran

10

Elnath

E ▸

ENE ▸

10°

September 5–6 • *Late in the evening the Moon passes Mars in the east.*

September 9–10 • *The Moon passes between Aldebaran and the Pleiades, shortly after midnight.*

Early morning 5:00 (BST)

Castor •

Pollux •

13

14

Venus

E ▸

10°

−25°

Early morning 6:00 (BST)

15

Regulus

16

Algieba •

E ▸

10°

September 13–14 • *The Moon passes Castor & Pollux and Venus.*

September 15–16 • *The crescent Moon passes Regulus and Algieba.*

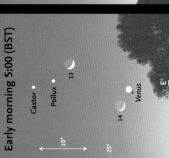

Evening 20:00 (BST)

Saturn Jupiter

25 24

Nunki

S ▸

10°

September 24–25 • *The Moon with Nunki, Jupiter and Saturn.*

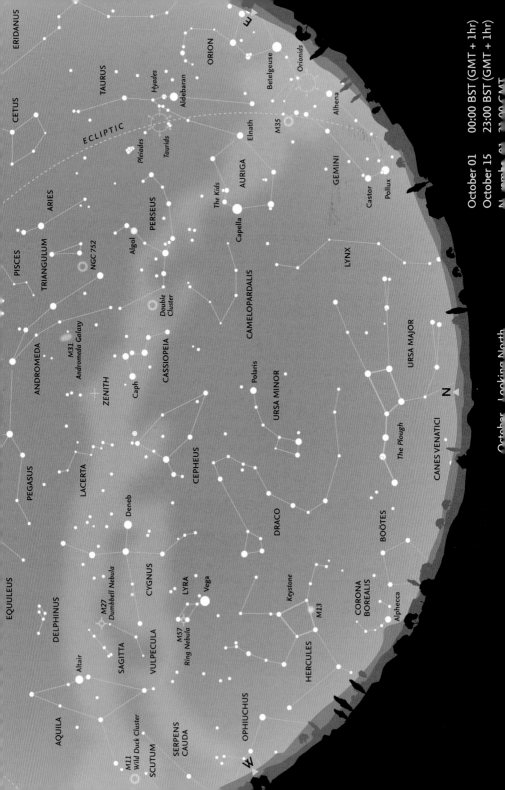

October 01   00:00 BST (GMT + 1hr)
October 15   23:00 BST (GMT + 1hr)

October - Looking North

October – Looking North

Ursa Major is grazing the horizon in the north, while high overhead are the constellations of **Cepheus, Cassiopeia** and **Perseus**, with the Milky Way between Cepheus and Cassiopeia at the zenith. **Auriga** is now clearly visible in the east, as is **Taurus** with the **Pleiades, Hyades** and orange **Aldebaran**. Also in the east, **Orion** and **Gemini** are starting to rise clear of the horizon.

The constellations of **Boötes** and **Corona Borealis** are now essentially lost to view in the northwest, and **Hercules** is also descending towards the western horizon. The three stars of the Summer Triangle are still clearly visible, although **Aquila** and **Altair** are beginning to approach the horizon in the west. Towards the end of the month (October 25) Summer Time ends in Europe, with Britain reverting to Greenwich Mean Time and Europe to Central European Time.

Meteors

The **Orionids** are the major, fairly reliable meteor shower active in October. Like the May **Eta Aquariid** shower, the Orionids are associated with Comet 1P/Halley. During this second pass through the stream of particles from the comet, slightly fewer meteors are seen than in May, but conditions are more favourable for northern observers. In both showers the meteors are very fast, and many leave persistent trains. Although the Orionid maximum is quoted as October 21–22, in fact there is a very broad maximum, lasting about a week from October 20 to 27, with hourly rates around 25. Occasionally rates are higher (50–70 per hour). In 2020, the broad maximum extends from three days before First Quarter, so conditions at maximum (October 21) are reasonably favourable, but then deteriorate.

The faint shower of the **Southern Taurids** (often with bright fireballs) peaks on October 10–11. The Southern Taurid maximum occurs when

The constellation of Perseus is not only the location of the radiant for the Perseid meteor shower, but is also well known for the pair of open clusters, known as the Double Cluster (near the top edge of the image) close to the border with Cassiopeia, and also for Algol (β Persei), a famous variable star that may be found at the centre of the photograph. The open star cluster, the Pleiades, in Taurus, is close to the bottom edge of the image.

the Moon is a waning crescent, so conditions are generally favourable. Towards the end of the month (around October 20), another shower (the **Northern Taurids**) begins to show activity, which peaks early in November. The parent comet for both Taurid showers is Comet 2P/Encke.

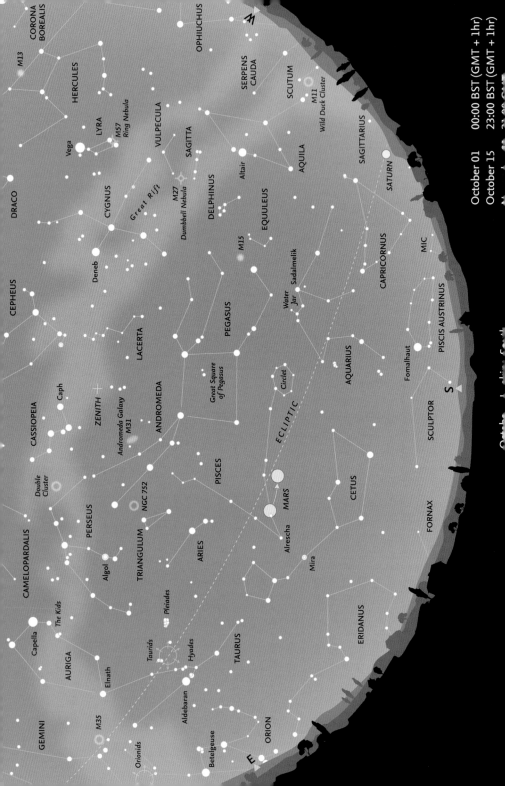

October – Looking South

The Great Square of **Pegasus** dominates the southern sky, framed by the two chains of stars that form the constellation of **Pisces**, together with **Alrescha** (α Piscium) at the point where the two lines of stars join. Also clearly visible is the constellation of **Cetus**, below Pegasus and Pisces. Although **Capricornus** is beginning to disappear, **Aquarius** to its east is well placed in the south, with solitary **Fomalhaut** and the constellation of **Piscis Austrinus** beneath it, close to the horizon.

The main band of the Milky Way and the Great Rift runs down from **Cygnus**, through **Vulpecula**, **Sagitta** and **Aquila** towards the western horizon. **Delphinus** and the tiny, unremarkable constellation of **Equuleus** lie between the band of the Milky Way and Pegasus. **Andromeda** is clearly visible high in the sky to the southeast, with the small constellation of **Triangulum** and the zodiacal constellation of **Aries** below it. **Perseus** is high in the east, and by now the **Pleiades** and **Taurus** are well clear of the horizon. Later in the night, and later in the month, **Orion** rises in the east, a sign that the autumn season has arrived and of the steady approach of winter.

The Moon's phases for October

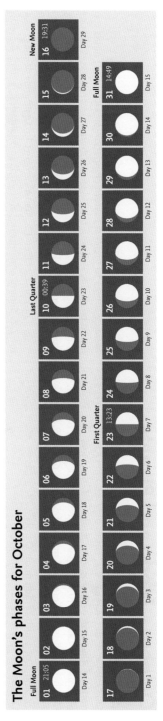

The constellation of Aquarius is one of the constellations that is visible in late summer and early autumn. The four stars forming the 'Y'-shape of the 'Water Jar' may be seen to the east of Sadalmelik (α Aquarii), the brightest star (top centre).

October – Moon and Planets

The Moon

On October 3, the Moon is 0.7° south of Mars. On October 7, it is 4.5° north of **Aldebaran** in **Taurus**. On October 13, the Moon is 4.5° north of **Regulus** in **Leo** and on October 13 it is also 4.4° north of **Venus**. By October 16, it is in **Virgo**, 7.0° north of **Spica**. It passes 5.7° north of **Antares** on October 19. The Moon is 2.0° south of **Jupiter** on October 22, and 3.0° south of **Saturn** on October 23. It passes **Mars** again on October 29, this time 3.0° to the south.

The planets

Mercury comes to greatest eastern elongation on October 1, but is too low to be readily visible. It then closes with the Sun and passes inferior conjunction on October 25. **Venus** begins the month in **Leo**. It is very close to **Regulus** (0.1° south) on October 3, but rapidly moves towards the Sun. **Mars** is retrograding in **Pisces** and comes to opposition (at mag. -2.6) on October 13. **Jupiter** remains in **Sagittarius** at mag. -2.4 to -2.2. **Saturn**, now moving eastwards in **Sagittarius**, is mag. 0.5–0.6. **Uranus** comes to opposition at mag. 5.7 in **Aries** on October 31. **Neptune** is mag. 7.8 in **Aquarius**.

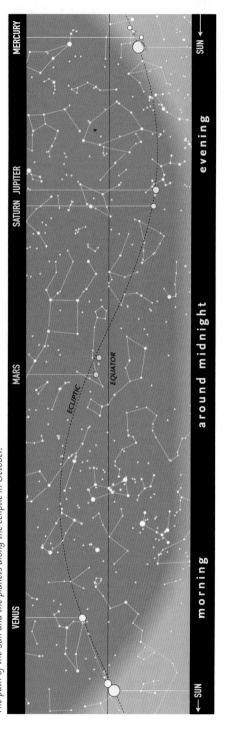

The path of the Sun and the planets along the ecliptic in October.

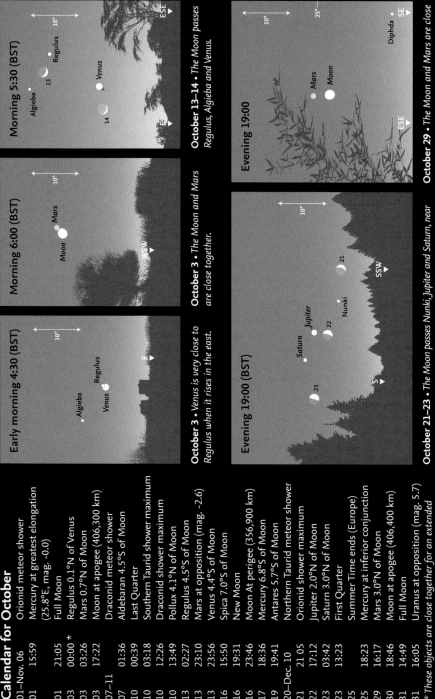

Calendar for October

Oct. 01–Nov. 06	15:59	Orionid meteor shower
01	21:05	Mercury at greatest elongation (25.8°E, mag. –0.0)
01	00:00 *	Full Moon
03	03:26	Regulus 0.1°N of Venus
03	17:22	Mars 0.7°N of Moon
07–11		Moon at apogee (406,300 km)
07	01:36	Draconid meteor shower
10	00:39	Aldebaran 4.5°S of Moon
10	03:18	Last Quarter
10	12:26	Southern Taurid shower maximum
10	13:49	Draconid shower maximum
13	02:27	Pollux 4.1°N of Moon
13	23:10	Regulus 4.5°S of Moon
13	23:56	Mars at opposition (mag. –2.6)
16	15:50	Venus 4.4°S of Moon
16	19:31	Spica 7.0°S of Moon
16	23:46	New Moon
17	18:36	Moon At perigee (356,900 km)
19	19:41	Mercury 6.8°S of Moon
20–Dec. 10		Antares 5.7°S of Moon
21	21 05	Northern Taurid meteor shower
22	17:12	Orionid shower maximum
23	03:42	Jupiter 2.0°N of Moon
23	13:23	Saturn 3.0°N of Moon
25		First Quarter
25	18:23	Summer Time ends (Europe)
29	16:17	Mercury at inferior conjunction
30	18:46	Mars 3.0°N of Moon
31	14:49	Moon at apogee (406,400 km)
31	16:05	Full Moon
		Uranus at opposition (mag. 5.7)

* These objects are close together for an extended period around this time.

Morning 5:30 (BST)

October 13–14 • The Moon passes Regulus, Algieba and Venus.

Early morning 4:30 (BST)

Morning 6:00 (BST)

October 3 • Venus is very close to Regulus when it rises in the east.

October 3 • The Moon and Mars are close together.

Evening 19:00 (BST)

October 29 • The Moon and Mars are close together, high in the east-southeast.

Evening 19:00 (BST)

October 21–23 • The Moon passes Nunki, Jupiter and Saturn, near the southern horizon.

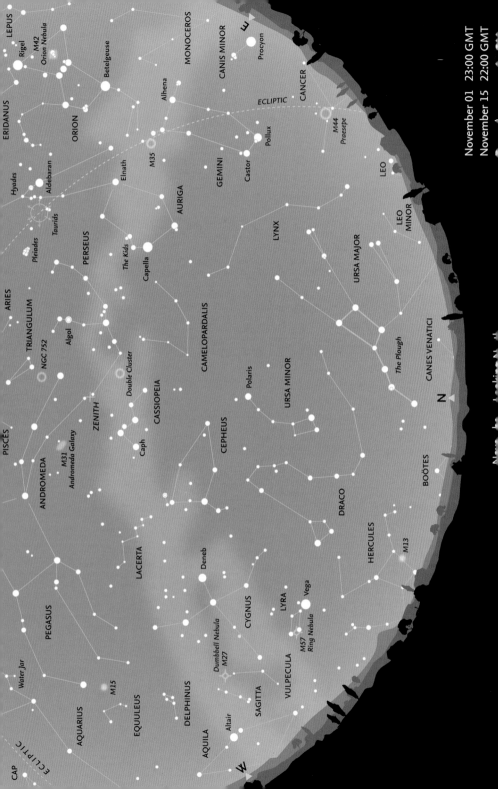

November – Looking North

Most of **Aquila** has now disappeared below the horizon, but two of the stars of the Summer Triangle, **Vega** in **Lyra** and **Deneb** in **Cygnus**, are still clearly visible in the west. The head of **Draco** is now low in the northwest and only a small portion of **Hercules** remains above the horizon. The southernmost stars of **Ursa Major** are now coming into view. The Milky Way arches overhead, with the denser star clouds in the west and the less heavily populated region through **Auriga** and **Monoceros** in the east. High overhead, **Cassiopeia** is near the zenith and **Cepheus** has swung round to the northwest, while Auriga is now high in the northeast. **Gemini**, with **Castor** and **Pollux**, is well clear of the eastern horizon, and even **Procyon** (α Canis Minoris) is just climbing into view almost due east.

At the beginning of the month (November 1) Daylight Saving Time comes to an end in North America.

Meteors

The **Northern Taurid** shower, which began in mid-October, reaches maximum – although with only a low rate of about five meteors per hour – on November 13–14. The Moon is a waning crescent, so conditions are moderately favourable. The shower gradually trails off, ending around December 10. There is an apparent 7-year periodicity in fireball activity, but 2020 is unlikely to be another peak year. Far more striking, however, are the **Leonids**, which have a relatively short period of activity (November 5–30), with maximum on November 17–18. This shower is associated with Comet 55P/Tempel-Tuttle and has shown extraordinary activity on various occasions, with many thousands of meteors per hour. High rates were seen in 1999, 2001 and 2002 (reaching about 3,000 meteors per hour) but have fallen dramatically since then. The rate in 2020 is likely to be about 15 per hour. These meteors are the fastest shower meteors recorded (about 70 km per second) and often leave persistent trains. The shower is very rich in faint meteors. In 2020, maximum is about two days after New Moon, so observing conditions are likely to be very favourable.

M31, the great Andromeda Galaxy, is about 220,000 light-years across, but has recently been found to be surrounded by an otherwise invisible halo of hot (10,000 to 100,000 degrees) ionized hydrogen at least five times that diameter, reaching halfway to our own Galaxy.

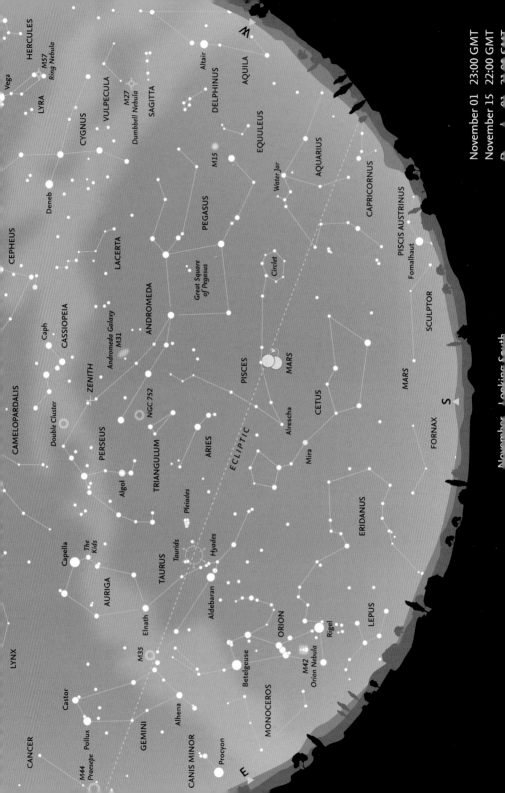

November. Looking South

November – Looking South

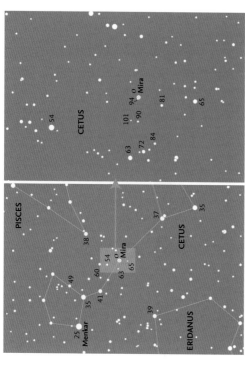

Orion has now risen above the eastern horizon, and part of the long, straggling constellation of **Eridanus** (which begins near **Rigel**) is visible to the west of Orion. Higher in the sky, **Taurus**, with the **Pleiades** cluster, and orange **Aldebaran** are now easy to observe. To their west, both **Pisces** and **Cetus** are close to the meridian. The famous long-period variable star, **Mira** (o Ceti), with a typical range of magnitude 3.4 to 9.8, is favourably placed for observation. In the southwest, **Capricornus** has slipped below the horizon, but **Aquarius** remains visible. Even farther west, **Altair** may be seen early in the night, but most of **Aquila** has already disappeared from view. **Delphinus**, together with **Sagitta** and **Vulpecula** in the Milky Way, will soon vanish for another year. Both **Pegasus** and **Andromeda** are easy to see, and one of the lines of stars that make up Andromeda finishes close to the zenith, which is also close to one of the outlying stars of **Perseus**, high in the east.

Finder and comparison charts for Mira (o Ceti). The chart on the left shows all stars brighter than magnitude 6.5. The chart on the right shows stars down to magnitude 10.0. The comparison star magnitudes are shown without the decimal point.

The Moon's phases for November

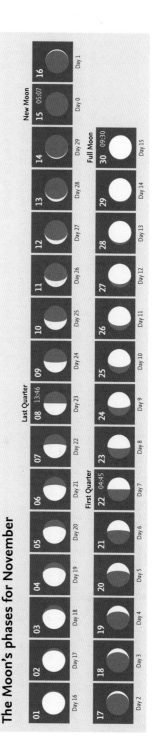

November – Moon and Planets

The Moon

On November 3, the Moon is 4.7° north of *Aldebaran* in *Taurus*. *Pollux* is one of the five bright stars that may be occulted by the Moon. On three occasions: November 6, December 4 and December 31, the Moon is 3.8° south of Pollux, the closest it comes in 2020. On October 9 it is 4.7° north of *Regulus* in *Leo*. On November 12 it passes 3.1° north of *Venus* in *Virgo* and the following day 7.0° north of *Spica*. On November 16, the Moon is 5.6° north of *Antares* in *Scorpius*. It passes 2.5° south of *Jupiter* and 2.9° south of *Saturn* on November 19. It is 4.9° south of *Mars* on November 25 and again 4.7° north of *Aldebaran* on November 30.

The planets

Mercury comes to greatest western elongation on November 10 (diagram page 22). *Venus* begins the month in the morning sky in *Virgo*, but (as in October) rapidly moves towards the Sun. *Mars* fades from mag. -2.1 to -1.2 over the month. It reverts to direct motion on November 17. *Jupiter* is in *Sagittarius*, moving eastwards at mag. -2.2 to -2.0, as is *Saturn* at mag. 0.6. *Uranus* is still retrograding in *Aries* at mag. 5.7. *Neptune* remains in *Aquarius* at mag. 7.8, resuming direct motion on November 30. On November 1, the minor planet (8) *Flora* is at opposition in Cetus at mag. 8.0.

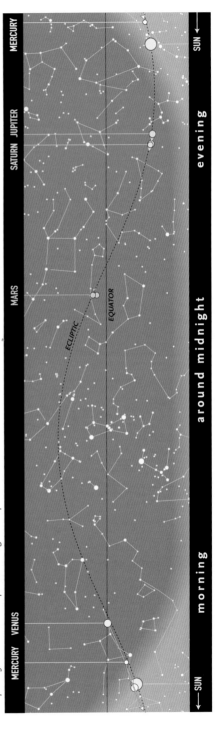

The path of the Sun and the planets along the ecliptic in November.

Calendar for November

01		Daylight Saving Time ends (North America)
01	06:50	Minor planet (8) Flora at opposition (mag. 8.0)
03	07:33	Aldebaran 4.7°S of Moon
05–29		Leonid meteor shower
06	20:24	Pollux 3.8°N of Moon
08	13:46	Last Quarter
09	10:48	Regulus 4.7°S of Moon
10	16:59	Mercury at greatest elongation (19.1°W, mag. −0.6)
12	21:30	Venus 3.1°S of Moon
13	02:52	Spica 7.0°S of Moon
13	12:05	Northern Taurid shower maximum
13	20:44	Mercury 1.7°S of Moon
14	11:48	Moon at perigee (357,800 km)
15	05:07	New Moon
15	13:00 ∗	Venus 4.1°N of Spica
16	06:30	Antares 5.6°S of Moon
17	10:53	Leonid shower maximum
19	08:56	Jupiter 2.5°N of Moon
19	14:51	Saturn 2.9°N of Moon
22	04:45	First Quarter
25	19:47	Mars 4.9°N of Moon
27	00:29	Moon at apogee (405,900 km)
30	09:30	Full Moon
30	09:43	Penumbral lunar eclipse
30	13:41	Aldebaran 4.7°S of Moon

∗ These objects are close together for an extended period around this time.

After midnight 1:00

Morning 6:30

November 9 • The Moon, Regulus and Algieba form an almost equilateral triangle, when they rise in the east.

November 12–14 • The narrow crescent Moon passes Venus, Spica and Mercury, shortly before sunrise.

Evening 17:00

Evening 21:00

Evening 18:00

November 18–19 • The Moon passes Nunki, Jupiter and Saturn.

November 25 • The Moon with Mars high in the southern sky.

November 29–30 • The Moon with Aldebaran and the Pleiades.

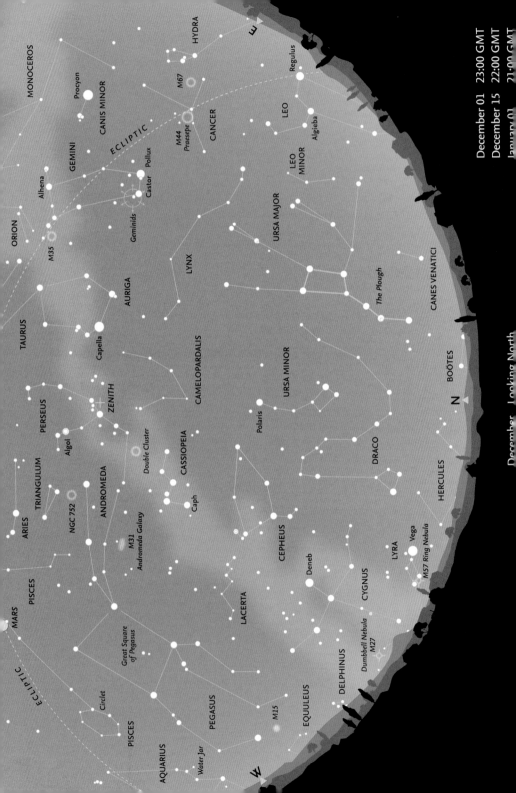

December   Looking North

December – Looking North

Ursa Major has now swung around and is starting to 'climb' in the east. The fainter stars in the southern part of the constellation are now fully in view. The other bear, ***Ursa Minor***, 'hangs' below ***Polaris*** in the north. Directly above it is the faint constellation of ***Camelopardalis***, with the other inconspicuous circumpolar constellation, ***Lynx***, to its east. ***Vega*** (α Lyrae) is skimming the horizon in the northwest, but ***Deneb*** (α Cygni) and most of ***Cygnus*** remain visible farther west. In the east, ***Regulus*** (α Leonis) and the constellation of ***Leo*** are beginning to rise above the horizon. ***Cancer*** stands high in the east, with ***Gemini*** even higher in the sky. ***Perseus*** is at the zenith, with ***Auriga*** and ***Capella*** between it and Gemini. Because it is so high in the sky, now is a good time to examine the star clouds of the fainter portion of the Milky Way, between ***Cassiopeia*** in the west to Gemini and ***Orion*** in the east.

Meteors

There is one significant meteor shower in December (the last major shower of the year). This is the ***Geminid*** shower, which is visible over the period December 3–16 and comes to maximum on December 13–14, at New Moon and the day after, so conditions are ideal. It is one of the most active showers of the year, and in some years is the strongest, with a peak rate of around 100 meteors per hour. It is the one major shower that shows good activity before midnight. The meteors have been found to have a much higher density than other meteors (which are derived from cometary material). It was eventually established that the Geminids and the asteroid (3200) Phaeton had similar orbits. So the Geminids are assumed to consist of denser, rocky material. They are slower than most other meteors and often appear to last longer. The brightest often break up into numerous luminous fragments that follow similar paths across the sky. There is a second shower: the ***Ursids***, active December 17–26, peaking on December 21–22, with rate

at maximum of 5–10, occasionally rising to 25 per hour. Maximum in 2020 is when the Moon is at First Quarter, so conditions are not very favourable.

The constellation of Cassiopeia is a familiar sight among the northern circumpolar constellations. It is always above the horizon, on the opposite side of Polaris (the North Star) to the equally well-known asterism of the seven stars of the Plough.

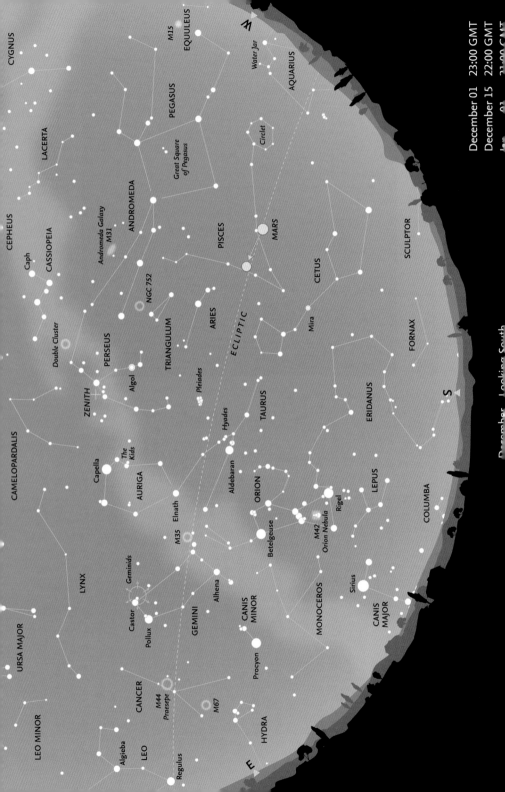

CYGNUS

EQUULEUS

M15

PEGASUS

LACERTA

Water Jar

AQUARIUS

CEPHEUS

Circlet

Great Square
of Pegasus

CASSIOPEIA

Caph

ANDROMEDA

Andromeda Galaxy
M31

PISCES

MARS

Double Cluster

NGC 752

PERSEUS

SCULPTOR

CETUS

CAMELOPARDALIS

ZENITH

TRIANGULUM

ARIES

Algol

Pleiades

ECLIPTIC

Mira

FORNAX

URSA MAJOR

The
Kids

Capella

Hyades

TAURUS

S

AURIGA

ERIDANUS

LYNX

Elnath

Aldebaran

M35

ORION

LEO MINOR

Geminids

Betelgeuse

M42
Orion Nebula

Rigel

LEPUS

Castor

Alhena

CANIS
MINOR

MONOCEROS

COLUMBA

Pollux

GEMINI

Sirius

CANCER

Procyon

CANIS
MAJOR

M44
Praesepe

M67

Algieba

LEO

HYDRA

Regulus

E

W

December 01   23:00 GMT
December 15   22:00 GMT

December Looking South

December – Looking South

The fine open cluster of the *Pleiades* is due south around 22:00, with the *Hyades* cluster, *Aldebaran* and the rest of *Taurus* clearly visible to the east. *Auriga* (with *Capella*) and *Gemini* (with *Castor* and *Pollux*) are both well-placed for observation. *Orion* has made a welcome return to the winter sky, and both *Canis Minor* (with *Procyon*) and *Canis Major* (with *Sirius*, the brightest star in the sky) are now well above the horizon. The small, poorly known constellation of *Lepus* lies to the south of Orion. In the west, *Aquarius* has now disappeared, and *Cetus* is becoming lower, but *Pisces* is still easily seen, as are the constellations of *Aries*, *Triangulum* and *Andromeda* above it. The Great Square of *Pegasus* is starting to plunge down towards the western horizon, and because of its orientation on the sky appears more like a large diamond, standing on one point, than a square.

The constellation of Taurus contains two contrasting open clusters: the compact Pleiades, with its striking blue-white stars, and the more scattered, 'V'-shaped Hyades, which are much closer to us. Orange Aldebaran (α Tauri) is not related to the Hyades, but lies between it and the Earth.

The Moon's phases for December

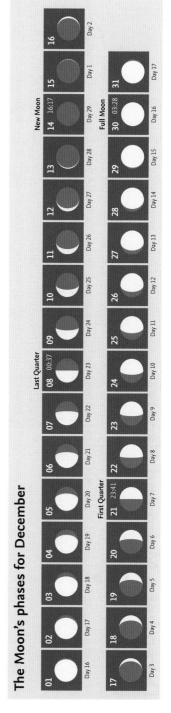

December – Moon and Planets

The Moon

On December 4, the Moon is 3.8° south of **Pollux** in **Gemini**. As on November 6 and December 31, this is the closest it comes to the star in 2020. On December 6, the Moon passes 4.8° north of **Regulus** in **Leo**. It is 7.1° north of **Spica** on December 10, and 0.8° north of **Venus** in **Libra** on December 12. The next day, December 13, it is 5.6° north of **Antares** in **Scorpius**. On December 17, the Moon passes 2.9° south of **Jupiter**, and less than one hour later, 3.1° south of **Saturn**. On December 23, it is 5.6° south of **Mars** in **Pisces**. The Moon passes 4.7° north of **Aldebaran** on December 27 and is again 3.8° south of **Pollux** on December 31.

The planets

Mercury is at superior conjunction on December 20. **Venus** (mag. -3.9) is a morning object in **Libra** for most of the month. **Mars**, moving eastwards in **Pisces**, fades fairly rapidly from mag. -1.1 to -0.3 over the month. **Jupiter** (mag. -2.0) is slowly overtaking **Saturn** (mag. 0.6) throughout the month, but the planets are closest (within 0.1°) on December 21, Saturn lying north of Jupiter. **Uranus** continues its slow retrograde motion in **Aries** at mag. 5.8, and **Neptune** (mag. 7.8) is also retrograding in **Aquarius**.

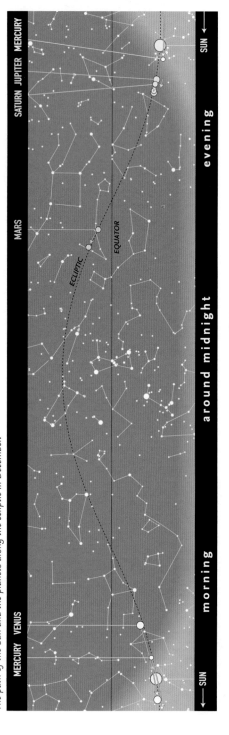

The path of the Sun and the planets along the ecliptic in December.

Calendar for December

03–16		Geminid meteor shower
04	01:56	Pollux 3.8°N of Moon
06	16:58	Regulus 4.8°S of Moon
08	00:37	Last Quarter
10	12:01	Spica 7.1°S of Moon
12	20:40	Venus 0.8°S of Moon
12	20:42	Moon at perigee (361,800 km)
13	00:49	Geminid shower maximum
13	17:32	Antares 5.6°S of Moon
14	10:33	Mercury 1.0°S of Moon
14	16:15	Total solar eclipse
14	16:17	New Moon
17–26		Ursid meteor shower
17	04:29	Jupiter 2.9°N of Moon
17	05:19	Saturn 3.1°N of Moon
20	03:26	Mercury at superior conjunction
21	10:02	Northern winter solstice
21	14:00 *	Saturn 0.1°N of Jupiter
21	23:41	First Quarter
22	09:11	Ursid meteor maximum
23	18:32	Mars 5.6°N of Moon
24	16:32	Moon at apogee (405,000 km)
27	20:54	Aldebaran 4.7°S of Moon
28–Jan. 12		Quadrantid meteor shower
30	03:28	Full Moon
31	08:13	Pollux 3.8°N of Moon

* *These objects are close together for an extended period around this time.*

Morning 6:00

10°

35°

10°

Castor
Pollux
Moon
Procyon
W

December 4 • *The Moon aligns with Castor and Pollux. Procyon is closer to the horizon.*

Morning 7:15

10°

12
Venus
13
SSE
SE
ESE

December 12–13 • *In the morning twilight, the narrow crescent Moon passes Venus in the southeast.*

Evening 16:45

10°

17
Saturn Jupiter
16
SSW
SW

December 16–17 • *Jupiter and Saturn are close when the Moon passes them.*

Evening 17:30

5°
12°

• Algedi (αCap)
Dabih (β Cap)
Saturn
Jupiter
WSW

1°
Saturn
Jupiter
5 x enlarged

December 21 • *Jupiter and Saturn are very close together. The actual conjunction (see calendar) occured in daylight.*

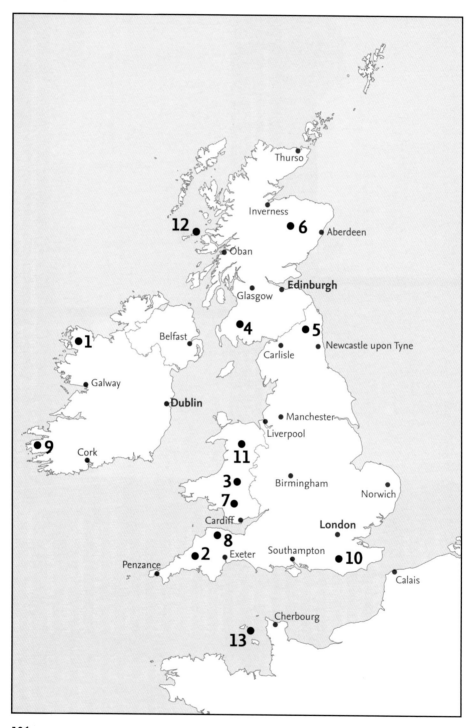

Thurso

Inverness ●6 ● Aberdeen

12 ●

● Oban

Edinburgh

Glasgow

●4 ●5

Belfast ● Carlisle ● Newcastle upon Tyne

●1

Galway

● Dublin

● Manchester

Liverpool

●9

Cork

11

●3 ● Birmingham

7● Norwich

Cardiff ● London

8 Exeter Southampton

●2 ● ●10

Penzance Calais

Cherbourg

13

Dark Sky Sites

International Dark-Sky Association Sites

The *International Dark-Sky Association* (IDA) recognizes various categories of sites that offer areas where the sky is dark at night, free from light pollution and particularly suitable for astonomical observing. A number of sites in Great Britain and Ireland have been given specific recognition and are shown on the map. These are:

1 *Ballycroy National Park and Wild Nephin Wilderness*

2 *Bodmin Moor Dark Sky Landscape*

3 *Elan Valley Estate*

4 *Galloway Forest Park*

5 *Northumberland National Park and Kielder Water & Forest Park*

6 *Tomintoul and Glenlivet – Cairngorms*

7 *Brecon Beacons National Park*

8 *Exmoor National Park*

9 *Kerry*

10 *Moore's Reserve – South Downs National Park*

11 *Snowdonia National Park*

12 *The island of Coll (Inner Hebrides, Scotland)*

13 *The island of Sark (Channel Islands)*

Details of these sites and web links may be found at the IDA website: https://www.darksky.org/ Many of these sites have major observatories or other facilities available for public observing (often at specific dates or times).

Dark Sky Discovery Sites

In Britain there is also the **Dark Sky Discovery** organization. This gives recognition to smaller sites, again free from immediate light pollution, that are open to observing at any time. Some sites are used for specific, public observing sessions. A full listing of sites is at:

https://www.darkskydiscovery.org.uk/

but specific events are publicized locally.

Glossary and Tables

aphelion	The point on an orbit that is farthest from the Sun.
apogee	The point on its orbit at which the Moon is farthest from the Earth.
appulse	The apparently close approach of two celestial objects; two planets, or a planet and star.
astronomical unit	(AU) The mean distance of the Earth from the Sun, 149,597,870 km.
celestial equator	The great circle on the celestial sphere that is in the same plane as the Earth's equator.
celestial sphere	The apparent sphere surrounding the Earth on which all celestial bodies (stars, planets, etc.) seem to be located.
conjunction	The point in time when two celestial objects have the same celestial longitude. In the case of the Sun and a planet, superior conjunction occurs when the planet lies on the far side of the Sun (as seen from Earth). For Mercury and Venus, inferior conjuction occurs when they pass between the Sun and the Earth.
direct motion	Motion from west to east on the sky.
ecliptic	The apparent path of the Sun across the sky throughout the year. Also: the plane of the Earth's orbit in space.
elongation	The point at which an inferior planet has the greatest angular distance from the Sun, as seen from Earth.
equinox	The two points during the year when night and day have equal duration. Also: the points on the sky at which the ecliptic intersects the celestial equator. The vernal (spring) equinox is of particular importance in astronomy.
gibbous	The stage in the sequence of phases at which the illumination of a body lies between half and full. In the case of the Moon, the term is applied to phases between First Quarter and Full, and between Full and Last Quarter.
inferior planet	Either of the planets Mercury or Venus, which have orbits inside that of the Earth.
magnitude	The brightness of a star, planet or other celestial body. It is a logarithmic scale, where larger numbers indicate fainter brightness. A difference of 5 in magnitude indicates a difference of 100 in actual brightness; thus, a first-magnitude star is 100 times as bright as one of sixth magnitude.
meridian	The great circle passing through the North and South Poles of a body and the observer's position; or the corresponding great circle on the celestial sphere that passes through the North and South Celestial Poles and also through the observer's zenith.
nadir	The point on the celestial sphere directly beneath the observer's feet, opposite the zenith.
occultation	The disappearance of one celestial body behind another, such as when stars or planets are hidden behind the Moon.
opposition	The point on a superior planet's orbit at which it is directly opposite the Sun in the sky.
perigee	The point on its orbit at which the Moon is closest to the Earth.
perihelion	The point on an orbit that is closest to the Sun.
retrograde motion	Motion from east to west on the sky.
superior planet	A planet that has an orbit outside that of the Earth.
vernal equinox	The point at which the Sun, in its apparent motion along the ecliptic, crosses the celestial equator from south to north. Also known as the First Point of Aries.
zenith	The point directly above the observer's head.
zodiac	A band, streching 8° on either side of the ecliptic, within which the Moon and planets appear to move. It consists of twelve equal areas, originally named after the constellation that once lay within it.

The Greek Alphabet

α	Alpha	ε	Epsilon	ι	Iota	ν	Nu	ρ	Rho	φ (φ)	Phi
β	Beta	ζ	Zeta	κ	Kappa	ξ	Xi	σ (ς)	Sigma	χ	Chi
γ	Gamma	η	Eta	λ	Lambda	ο	Omicron	τ	Tau	ψ	Psi
δ	Delta	θ (ϑ)	Theta	μ	Mu	π	Pi	υ	Upsilon	ω	Omega

The Constellations

There are 88 constellations covering the whole of the celestial sphere, but 24 of these in the southern hemisphere can never be seen (even in part) from the latitude of Britain and Ireland, so are omitted from this table. The names themselves are expressed in Latin, and the names of stars are frequently given by Greek letters followed by the genitive of the constellation name. The genitives and English names of the various constellations are included.

Name	Genitive	Abbr.	English name
Andromeda	Andromeda	And	Andromeda
Antlia	Antliae	Ant	Air Pump
Aquarius	Aquarii	Aqr	Water Bearer
Aquila	Aquilae	Aql	Eagle
Aries	Arietis	Ari	Ram
Auriga	Aurigae	Aur	Charioteer
Boötes	Boötis	Boo	Herdsman
Camelopardalis	Camelopardalis	Cam	Giraffe
Cancer	Cancri	Cnc	Crab
Canes Venatici	Canum Venaticorum	CVn	Hunting Dogs
Canis Major	Canis Majoris	CMa	Big Dog
Canis Minor	Canis Minoris	CMi	Little Dog
Capricornus	Capricorni	Cap	Sea Goat
Cassiopeia	Cassiopeiae	Cas	Cassiopeia
Centaurus	Centauri	Cen	Centaur
Cepheus	Cephei	Cep	Cepheus
Cetus	Ceti	Cet	Whale
Columba	Columbae	Col	Dove
Coma Berenices	Comae Berenices	Com	Berenice's Hair
Corona Australis	Coronae Australis	CrA	Southern Crown
Corona Borealis	Coronae Borealis	CrB	Northern Crown
Corvus	Corvi	Crv	Crow
Crater	Crateris	Crt	Cup
Cygnus	Cygni	Cyg	Swan
Delphinus	Delphini	Del	Dolphin
Draco	Draconis	Dra	Dragon
Equuleus	Equulei	Equ	Little Horse
Eridanus	Eridani	Eri	River Eridanus
Fornax	Fornacis	For	Furnace
Gemini	Geminorum	Gem	Twins
Hercules	Herculis	Her	Hercules
Hydra	Hydrae	Hya	Water Snake

Name	Genitive	Abbr.	English name
Lacerta	Lacertae	Lac	Lizard
Leo	Leonis	Leo	Lion
Leo Minor	Leonis Minoris	LMi	Little Lion
Lepus	Leporis	Lep	Hare
Libra	Librae	Lib	Scales
Lupus	Lupi	Lup	Wolf
Lynx	Lyncis	Lyn	Lynx
Lyra	Lyrae	Lyr	Lyre
Microscopium	Microscopii	Mic	Microscope
Monoceros	Monocerotis	Mon	Unicorn
Ophiuchus	Ophiuchi	Oph	Serpent Bearer
Orion	Orionis	Ori	Orion
Pegasus	Pegasi	Peg	Pegasus
Perseus	Persei	Per	Perseus
Pisces	Piscium	Psc	Fishes
Piscis Austrinus	Piscis Austrini	PsA	Southern Fish
Puppis	Puppis	Pup	Stern
Pyxis	Pyxidis	Pyx	Compass
Sagitta	Sagittae	Sge	Arrow
Sagittarius	Sagittarii	Sgr	Archer
Scorpius	Scorpii	Sco	Scorpion
Sculptor	Sculptoris	Scl	Sculptor
Scutum	Scuti	Sct	Shield
Serpens	Serpentis	Ser	Serpent
Sextans	Sextantis	Sex	Sextant
Taurus	Tauri	Tau	Bull
Triangulum	Trianguli	Tri	Triangle
Ursa Major	Ursae Majoris	UMa	Great Bear
Ursa Minor	Ursae Minoris	UMi	Lesser Bear
Vela	Velorum	Vel	Sails
Virgo	Virginis	Vir	Virgin
Vulpecula	Vulpeculae	Vul	Fox

Some common asterisms

Belt of Orion	δ, ε and ζ Orionis
Big Dipper	α, β, γ, δ, ε, ζ and η Ursae Majoris
Circlet	γ, θ, ι, λ and κ Piscium
Guards (or Guardians)	β and γ Ursae Minoris
Head of Cetus	α, γ, ξ², μ and λ Ceti
Head of Draco	β, γ, ξ and ν Draconis
Head of Hydra	δ, ε, ζ, η, ρ and σ Hydrae
Keystone	ε, ζ, η and π Herculis
Kids	ε, ζ and η Aurigae
Little Dipper	β, γ, η, ζ, ε, δ and α Ursae Minoris
Lozenge	= Head of Draco
Milk Dipper	ζ, γ, σ, φ and λ Sagittarii
Plough or Big Dipper	α, β, γ, δ, ε, ζ and η Ursae Majoris
Pointers	α and β Ursae Majoris
Sickle	α, η, γ, ζ, μ and ε Leonis
Square of Pegasus	α, β and γ Pegasi with α Andromedae
Sword of Orion	θ and ι Orionis
Teapot	γ, ε, δ, λ, φ, σ, τ and ζ Sagittarii
Wain (or Charles' Wain)	= Plough
Water Jar	γ, η, κ and ζ Aquarii
Y of Aquarius	= Water Jar

Acknowledgements

Steve Edberg, La Cañada, California: p.21 (Annular eclipse); all individual constellation photographs

NASA: p.21 (Solar eclipse paths)

Damian Peach, Hamble, Hants.: p.29 (Comet Lovejoy)

peresanz/Shutterstock: p.35 (Orion)

Denis Buczyncki, Portmahomack, Ross-shire: p.29 (Comet Iwamoto); pp.32 & 37 (Quadrantid fireball); p.65 (Noctilucent clouds)

Jens Hackmann, Bad Mergentheim, Germany: p.77 (Perseid fireball)

Ken Sperber, California: p.83 (Double Cluster)

Adam Evans CC by 2.0: p.95 (Andromeda Galaxy)

Specialist editorial support was provided by Dr Emily Drabek-Maunder, Public Astronomy Officer at the Royal Observatory, part of Royal Museums Greenwich.

Further Information

Books

Bone, Neil (1993), *Observer's Handbook: Meteors*, George Philip, London & Sky Publ. Corp., Cambridge, Mass.

Cook, J., ed. (1999), *The Hatfield Photographic Lunar Atlas*, Springer-Verlag, New York

Dunlop, Storm (1999), *Wild Guide to the Night Sky*, HarperCollins, London

Dunlop, Storm (2012), *Practical Astronomy*, 3rd edn, Philip's, London

Dunlop, Storm, Rükl, Antonin & Tirion, Wil (2005), *Collins Atlas of the Night Sky*, HarperCollins, London

Heifetz, Milton & Tirion, Wil (2017), *A Walk through the Heavens: A Guide to Stars and Consellations and their Legends*, 4th edition, Cambridge University Press, Cambridge

O'Meara, Stephen J. (2008), *Observing the Night Sky with Binoculars*, Cambridge University Press, Cambridge

Ridpath, Ian, ed. (2004), *Norton's Star Atlas*, 20th edn, Pi Press, New York

Ridpath, Ian, ed. (2003), *Oxford Dictionary of Astronomy*, 2nd edn, Oxford University Press, Oxford

Ridpath, Ian & Tirion, Wil (2004), *Collins Gem - Stars*, HarperCollins, London

Ridpath, Ian & Tirion, Wil (2011), *Collins Pocket Guide Stars and Planets*, 4th edn, HarperCollins, London

Ridpath, Ian & Tirion, Wil (2012), *Monthly Sky Guide*, 9th edn, Cambridge University Press

Rükl, Antonín (1990), *Hamlyn Atlas of the Moon*, Hamlyn, London & Astro Media Inc., Milwaukee

Rükl, Antonín (2004), *Atlas of the Moon*, Sky Publishing Corp., Cambridge, Mass.

Scagell, Robin (2000), *Philip's Stargazing with a Telescope*, George Philip, London

Tirion, Wil (2011), *Cambridge Star Atlas*, 4th edn, Cambridge University Press, Cambridge

Tirion, Wil & Sinnott, Roger (1999), *Sky Atlas 2000.0*, 2nd edn, Sky Publishing Corp., Cambridge, Mass. & Cambridge University Press, Cambridge

Journals

Astronomy, Astro Media Corp., 21027 Crossroads Circle, P.O. Box 1612, Waukesha, WI 53187-1612 USA. http://www.astronomy.com

Astronomy Now, Pole Star Publications, PO Box 175, Tonbridge, Kent TN10 4QX UK. http://www.astronomynow.com

Sky at Night Magazine, BBC publications, London. http://skyatnightmagazine.com

Sky & Telescope, Sky Publishing Corp., Cambridge, MA 02138-1200 USA. http://www.skyandtelescope.com/

Societies

British Astronomical Association, Burlington House, Piccadilly, London W1J 0DU. http://www.britastro.org/

The principal British organization for amateur astronomers (with some professional members), particularly for those interested in carrying out observational programmes. Its membership is, however, worldwide. It publishes fully refereed, scientific papers and other material in its well-regarded journal.

Federation of Astronomical Societies, Secretary: Ken Sheldon, Whitehaven, Maytree Road, Lower Moor, Pershore, Worcs. WR10 2NY. http://www.fedastro.org.uk/fas/

An organization that is able to provide contact information for local astronomical societies in the United Kingdom.

Royal Astronomical Society, Burlington House, Piccadilly, London W1J 0BQ. http://www.ras.org.uk/

The premier astronomical society, with membership primarily drawn from professionals and experienced amateurs. It has an exceptional library and is a designated centre for the retention of certain classes of astronomical data. Its publications are the standard medium for dissemination of astronomical research.

Society for Popular Astronomy, 36 Fairway, Keyworth, Nottingham NG12 5DU.
http://www.popastro.com/
A society for astronomical beginners of all ages, which concentrates on increasing members' understanding and enjoyment, but which does have some observational programmes. Its journal is entitled *Popular Astronomy*.

Software

Planetary, Stellar and Lunar Visibility, (planetary and eclipse freeware): Alcyone Software, Germany.
http://www.alcyone.de

Redshift, Redshift-Live. http://www.redshift-live.com/en/

Starry Night & Starry Night Pro, Sienna Software Inc., Toronto, Canada. http://www.starrynight.com

Internet sources

There are numerous sites with information about all aspects of astronomy, and all of those have numerous links. Although many amateur sites are excellent, treat any statements and data with caution. The sites listed below offer accurate information. Please note that the URLs may change. If so, use a good search engine, such as Google, to locate the information source.

Information

Astronomical data (inc. eclipses) HM Nautical Almanac Office: http://astro.ukho.gov.uk

Auroral information Michigan Tech: http://www.geo.mtu.edu/weather/aurora/

Comets JPL Solar System Dynamics: http://ssd.jpl.nasa.gov/

American Meteor Society: http://amsmeteors.org/

Deep-sky objects Saguaro Astronomy Club Database: http://www.virtualcolony.com/sac/

Eclipses: NASA Eclipse Page: http://eclipse.gsfc.nasa.gov/eclipse.html

Moon (inc. Atlas) Inconstant Moon: http://www.inconstantmoon.com/

Planets Planetary Fact Sheets: http://nssdc.gsfc.nasa.gov/planetary/planetfact.html

Satellites (inc. International Space Station)
Heavens Above: http://www.heavens-above.com/
Visual Satellite Observer: http://www.satobs.org/

Star Chart: http://www.skyandtelescope.com/observing/interactive-sky-watching-tools/interactive-sky-chart/

What's Visible
Skyhound: http://www.skyhound.com/sh/skyhound.html
Skyview Cafe: http://www.skyviewcafe.com

Institutes and Organizations

European Space Agency: http://www.esa.int/

International Dark-Sky Association: http://www.darksky.org/

Jet Propulsion Laboratory: http://www.jpl.nasa.gov/

Lunar and Planetary Institute: http://www.lpi.usra.edu/

National Aeronautics and Space Administration: http://www.hq.nasa.gov/

Solar Data Analysis Center: http://umbra.gsfc.nasa.gov/

Space Telescope Science Institute: http://www.stsci.edu/